Handbook of Soil and Soilless Media Under Protected Cultivation

Handbook of

Soil and Soilless Media Under Protected Cultivation

Contributors :
Noel Ortuño,
José Antonio Castillo, *et al.*

Volume 2

KOROS PRESS LIMITED
London, UK

Handbook of Soil and Soilless Media Under Protected Cultivation
Contributors : Noel Ortuño *and* José Antonio Castillo, *et al.*

Published by Koros Press Limited

www.korospress.com

United Kingdom

Printed in 2017 for Sale in the Indian Subcontinent

Handbook of Soil and Soilless Media Under Protected Cultivation

ISBN: 978-1-78163-472-1

British Library Cataloguing in Publication Data
A CIP record for this book is available from the British Library

Exclusively distributed by CBS Publishers & Distributors Pvt. Ltd.

Sales & Distribution Rights only for India, Pakistan, Bangladesh, Sri Lanka, Nepal and Bhutan.This book is not to be sold outside these territories.

PREFACE

There may be as many definitions of soil as there are people who worry about it. One way to think about it is to define what it is not. Dirt is not soil. It may have once been soil, but once it was removed from its place in the landscape, it probably ceased being soil. It can become soil again in the right circumstances. Potting soils are not soils, they are growth media. Pure sand or water can serve as media, and neither can be confused with soils. This discussion will concentrate on soils. Media properties are covered elsewhere. For wholistic purposes, soil should be thought of as a three-dimensional living, dynamic resource that supports plant life. It has constantly changing biological, chemical, and physical properties that affect its ability to function in the environment. Let us look at soils in the reductionistic manner. If we break a soil down into its components we will find different sized mineral particles, organic matter, some things that are currently alive, and some air and water. The different size mineral particles will break out into three distinct classes: The biggest ones are called sand. They will have a diameter of 0.05 to 2.0 mm. You can pick most sand particles out with the naked eye. Intermediate sized particles are called silt. It requires 10x to 50x magnification to distinguish them by visual examination. The diameter ranges from 0.002 to 0.05 mm. The smallest particles are called clay. They will have diameters of less than 0.002 mm and required magnification greater than 50x to see. Note that clay particles have diameters ranging from 25 times less than the smallest sand or largest silt to 1000 times less than the largest sand particles. Organic matter is the residue left after multitudes of soil microbe generations have used the carbon and other easily usable constituents of once-living flora and fauna that ended up in the soil matrix. True organic matter, or humic substances, are relatively stable and resist any more ërapid' breakdown, although there will continue to be some activity on it. There are astonishing numbers of living organisms in most soils. They are continuously recycling nutrients in the soil released from the minerals, added plant material, added water, and previous generations of the organisms.

Under protected cultivation nutrient availability in the root medium is usually kept high because of the great nutrient absorption by growing crops. Most of the fertilizer requirements are usually supplied by irrigation water which is often given by sprinkler and trickle systems of irrigation. Base dressing of phosphorus and potassium are supplied under certain conditions. In spite of the fact that plant development depends mainly on nutrient supplied with irrigation water, both of those nutrients added or found initially in the soil undergo various chemical reactions depending on the conditions prevailing in the soil. The soil factors that may affect the nutrient availability to plants are: soil texture, pH, calcium carbonate content, E.C., organic matter, temperature and moisture content. Nitrogen fertigation has received the greatest attention and is applied in the forms of ammonium, nitrate and urea. These forms of nitrogen interact differently with soil. In calcareous and alkaline soils, the use of ammonium has sometimes considerable advantages over the use of nitrate as the pH drops in the rhizosphere. N_2 and K achieve good movement and diffusion, they become more available for absorption in the active root zone. On the other hand, P mobility in the soil is more limited but improves under fertigation. Micronutrient deficiencies are common in the newly reclaimed sandy and calcareous soils, some alluvial soils are considered marginal with respect to Zn and Fe availability.

Contents

This page left intentionally blank.

LIST OF CONTRIBUTORS

Hassan S. Al-Zahrani

King Abdulaziz University, Faculty of Science, Department of Biology, P. O. Box 80054, Jeddah 21589, Saudi Arabia

Noel Ortuño

Fundación PROINPA, Av. Meneces Km 4, El Paso, Cochabamba, Bolivia; E-Mails: m.claros@proinpa.org (M.C.); o.navia@proinpa.org (O.N.); m.angulo @proinpa.org (M.A.); d.barja@proinpa.org (D.B.); c.gutierrez@proinpa.org (C.G.); v.angulo@proinpa.org (V.A.)

José Antonio Castillo

Fundación PROINPA, Av. Meneces Km 4, El Paso, Cochabamba, Bolivia; E-Mails: m.claros@proinpa.org (M.C.); o.navia@proinpa.org (O.N.); m.angulo @proinpa.org (M.A.); d.barja@proinpa.org (D.B.); c.gutierrez@proinpa.org (C.G.); v.angulo@proinpa.org (V.A.)

Mayra Claros

Fundación PROINPA, Av. Meneces Km 4, El Paso, Cochabamba, Bolivia; E-Mails: m.claros@proinpa.org (M.C.); o.navia@proinpa.org (O.N.); m.angulo @proinpa.org (M.A.); d.barja@proinpa.org (D.B.); c.gutierrez@proinpa.org (C.G.); v.angulo@proinpa.org (V.A.)

Oscar Navia

Fundación PROINPA, Av. Meneces Km 4, El Paso, Cochabamba, Bolivia; E-Mails: m.claros@proinpa.org (M.C.); o.navia@proinpa.org (O.N.); m.angulo @proinpa.org (M.A.); d.barja@proinpa.org (D.B.); c.gutierrez@proinpa.org (C.G.); v.angulo@proinpa.org (V.A.)

Marlene Angulo

Fundación PROINPA, Av. Meneces Km 4, El Paso, Cochabamba, Bolivia; E-Mails: m.claros@proinpa.org (M.C.); o.navia@proinpa.org (O.N.); m.angulo @proinpa.org (M.A.); d.barja@proinpa.org (D.B.); c.gutierrez@proinpa.org (C.G.); v.angulo@proinpa.org (V.A.)

Daniel Barja

Fundación PROINPA, Av. Meneces Km 4, El Paso, Cochabamba, Bolivia; E-Mails: m.claros@proinpa.org (M.C.); o.navia@proinpa.org (O.N.); m.angulo @proinpa.org (M.A.); d.barja@proinpa.org (D.B.); c.gutierrez@proinpa.org (C.G.); v.angulo@proinpa.org (V.A.)

Claudia Gutiérrez

Fundación PROINPA, Av. Meneces Km 4, El Paso, Cochabamba, Bolivia; E-Mails: m.claros@proinpa.org (M.C.); o.navia@proinpa.org (O.N.); m.angulo @proinpa.org (M.A.); d.barja@proinpa.org (D.B.); c.gutierrez@proinpa.org (C.G.); v.angulo@proinpa.org (V.A.)

Violeta Angulo

Fundación PROINPA, Av. Meneces Km 4, El Paso, Cochabamba, Bolivia; E-Mails: m.claros@proinpa.org (M.C.); o.navia@proinpa.org (O.N.); m.angulo @proinpa.org (M.A.); d.barja@proinpa.org (D.B.); c.gutierrez@proinpa.org (C.G.); v.angulo@proinpa.org (V.A.)

Chapter 14

GLOBAL HYDROLITIC ACTIVITY (FDA TEST)

DEFINITION

The term "test article" is found in the FDA regulations on Protection of Human Subjects (21 CFR 50.3, Definitions (j)). The term includes drugs (including botanicals, biologicals, and gene therapy, and genetically derived products that meet the definition of a "drug"), and medical devices for human use. The FDA has statutory authority to regulate the development and marketing of these products.

KINDS

What kind of review process will the IRB conduct if the intent of the research in humans is to develop information about safety and efficacy of drugs or devices for submission to, or inspection by, the FDA?

Type of IRB

If a test article does not meet the criteria for an expedited review procedure, the study must be reviewed at a convened IRB meeting.

Drugs, Botanicals, Biologicals, or Other Substances

JHM IRB review of research involving unlicensed or unapproved drugs, botanicals, biologicals, or other substances (*e.g.*, food additives, food derivatives, vitamins, minerals, extracts, etc.) must be conducted at a convened meeting. The review must include the presence of a P&T IRB member who will make a recommendation concerning the risks and proposed benefits of each product to be used in the study. The IRB review will address the need for submission of an FDA Investigational New Drug Application (IND) prior to the use of the test article in the research study. The IRB may determine that an IND is not required if all the conditions set forth in 21 CFR 312.2(b) have been met. The IRB will determine whether to require submission of an IND, or if appropriate, to require the Primary Investigator (PI) to seek clarification from the FDA about the need for obtaining an IND before the IRB completes its review, and before the research may begin. Consent forms for IND studies must clearly state that the "test article" is investigational or experimental.

For all studies that involve a drug product, the PI must complete the eIRB section on Drugs. The PI must clarify for the IRB whether or not there is an FDA issued IND number for the use of the test article in the protocol, and if not, why not. Appropriate documentation (investigator drug brochure, package insert, FDA issued IND letter when the PI or other study team member holds the IND, etc.) should be included in the eIRB application.

Medical Devices

When the test article is a medical device, the JHM IRB reviewing the investigational medical device study will determine if the application will involve : a device approved by the FDA for marketing, or an investigational use of an unapproved device.

For unapproved devices, the IRB will determine :

A. Whether the test article requires an Investigational Device Exemption (IDE) from the FDA or is exempt from the IDE requirements.
   - The JHM IRB may consult the FDA or ask the Principal Investigator (PI) to consult the FDA for an opinion for difficult device assessments.

B. Whether the device poses a "Significant Risk" (SR) to the participants in the study.
   - The risk determination made by the IRB will be based upon both the nature of the device AND the proposed use of the device in the study as outlined in 21 CFR 812(3)(m)(1-4).
   - If the device is SR, the investigator must provide documentation from the FDA that the device either :
     i. Has a 510(k) designation or an Investigational Device Exemption (IDE) before the IRB may take final action to approve a protocol; or
     ii. If the JHM IRB believes that a device is SR, and the sponsor believes it to be NSR, the JHM IRB will require the PI to notify the sponsor of the JHM IRB determination. The sponsor may then consult with the FDA for an agency opinion. The FDA opinion must be provided to the IRB. If the sponsor chooses not to consult with the FDA and/or not to submit an IDE application to the FDA, the protocol will not be approved at JHM.

C. Whether the device poses a "Non-Significant Risk" (NSR) to the participants in the study.

- In the case of an NSR claim, the initial assessment of whether or not the device study is NSR is made by the sponsor/investigator and the justification for this assessment is submitted in the eIRB application. The IRB then determines whether that assessment is appropriate. If it agrees, the IRB will record the rationale for its NSR assessment on the review sheet for determinations and in the minutes for a convened meeting. All NSR devices under 21 CFR 812.2(b) are "considered to have approved applications for IDEs".
- The IRB will consider whether some NSR devices meet the criteria for exempt classification as a diagnostic device under 21 CFR 812(c)(3). If an investigational device meets those criteria, the IRB will document that it is exempt from the IDE requirements.

Consent forms for IDE studies must clearly state that the "test article" is investigational or experimental.

Institutional Requirement for Use of Medical Devices at Some JH Locations

Devices that are either electrical or battery powered and would be used at the East Baltimore JH Hospital and/or JH Bayview Medical Center must be reviewed and approved by Clinical Engineering at that location prior to use. Documentation of Clinical Engineering approval must be uploaded into the application and submitted to the JHM IRB for review. If the electrical or battery operated device would be used at both locations, then approval documents from both Clinical Engineering Departments must be uploaded into the application and submitted to the JHM IRB for review.

What are the Responsibilities of Investigators Who are also the Sponsor of IND or IDE Clinical Investigations

Devices that are either elecJHM investigators who submit protocols involving FDA test articles and for which they hold the IND or IDE are sponsor-investigators must adhere to the same regulatory responsibilities as any other sponsor. The application must include all supporting FDA documentation.

Additionally, in cases where an IND product will be manufactured at Hopkins, the PI must submit documentation that the proposed drug preparation meets standards for current Good Manufacturing Practice.

INSTITUTIONAL REQUIREMENT FOR SPONSOR-INVESTIGATORS

IND Studies

FDA test articles and for which they hold the IND or IDE are sponsor-investigators must adhere to the same regulatory responsibilities as any oa. All research studies for which an investigator holds an IND must comply with the Drug Use for Clinical Investigation (DUCI) policy "Organisation Policy on Drug Use and Control in Clinical Investigations (Policy No. 103.19(a)) ". The Investigational Drug Service (IDS) will monitor the study for compliance with sponsor-investigator requirements.

IDE Studies

- The Organisation requires a monitoring process for receipt, dispensing, and record keeping concerning devices that are studied using an IDE granted by FDA. The monitoring process shall be performed by the Compliance Monitoring Specialists of the OHSR.
- Researchers who serve as a sponsor-investigator for an IDE research study are required by the Organisation to follow FDA regulations 21 CFR 812 Subpart C applicable to sponsor responsibilities. Compliance Monitoring Specialists of the OHSR will conduct a monitoring visit for investigators holding an IDE to determine compliance with FDA sponsor requirements in 21 CFR 812 before initiation of the research.

FDA REPORTING AND RECORD KEEPING REQUIREMENTS FOR IND AND IDE RESEARCH STUDIES

Sponsor-investigators must comply with the record-keeping and reporting requirements provided in the Federal Regulations for INDs and for IDEs. In addition, the sponsor-investigator must comply with institutional record-keeping requirements and unanticipated problem reporting.

See : JHM IRB Guidance on Data Retention; and the "Organisation Policy on Reports of Unanticipated Problems Involving Risks to Participants or Others :

- A sponsor-investigator holding an IDE must report to the FDA and all reviewing IRBs and participating investigators any evaluation of an unanticipated device effect within ten (10) working days of first receiving notice of the effect.
- A sponsor-investigator holding an IND must satisfy the requirement to ensure that the FDA "and all participating investigators are promptly informed of significant new adverse effects or risks with respect to the drug".

FDA ANNUAL REPORTS REQUIRED FOR SPONSOR-INVESTIGATORS OF INDS

Each year, within 60 days of the anniversary date that the IND went into effect, the sponsor-investigator must submit to the FDA a brief progress report that includes the following information for each protocol actively under the IND.

Individual Study Information

- The title of the study, its purpose, a statement identifying the patient population, and a statement as to whether the study is completed;
- The total number of subjects initially planned for inclusion; the number entered into the study to date tabulated by age group, gender and race; the number whose participation in the study was completed as planned; and the number of subjects who dropped out of the study for any reason; and
- If the study has been completed, or if interim results are known, a brief description of available study results.

Summary Information

- A narrative or tabular summary showing the most frequent and most serious adverse experiences by body system;
- A summary of all IND safety reports submitted during the past year;
- A list of subjects who died during participation in the study, with the cause of death for each subject;
- A list of subjects who dropped out during the course of the investigation in association with any adverse experience, whether or not drug related;
- A brief description of what, if anything, was obtained that is pertinent to an understanding of the drug's actions, including for example, information about dose response, information about controlled trials, and information about bioavailability;
- A list of preclinical studies (including animal studies) completed or in progress during the past year and a summary of the major preclinical findings; and
- A summary of any significant manufacturing or microbiological changes made during the past year.

General Investigational Plan

A description of the general plan for the upcoming year to replace that which was submitted in the previous year.

This plan should include all information required in 21 CFR 312.23(a)(3)(iv) :

- Revised investigator brochure and description of the revision, if applicable.
- Description of any Phase 1 protocol modifications made during the year and not previously reported to the IND in a protocol amendment.
- Brief summary of significant foreign marketing developments with the drug during the past year, if applicable.
- If desired, a log of any outstanding business with respect to the IND for which the sponsor/investigator requests or expects a reply, comment, or meeting.

FDA Required Reports for Sponsor-Investigators of IDEs.

A SPONSOR-INVESTIGATOR OF AN IDE MUST SUBMIT THE FOLLOWING REPORTS

Progress Reports

- A sponsor-investigator shall submit progress reports on the investigation at regular intervals, and at least yearly, to all reviewing IRBs.
- In the case of a significant risk (SR) device, the sponsor shall also submit progress reports to the FDA.
- A sponsor of a treatment IDE shall submit semi-annual progress reports to all reviewing IRBs and FDA in accordance with 21 CFR 812.36(f).

Recall and Device Disposition

A sponsor shall notify FDA and all reviewing IRBs of any request that an investigator return, repair, or otherwise dispose of any units of a device. Such notice shall occur within 30 working days after the request is made, and shall state why the request was made.

Final Report

- In the case of an SR device, the sponsor shall notify FDA within 30 working days of the completion or termination of the investigation and shall submit a final report to FDA and all reviewing IRBs and participating investigators within 6 months of the completion or termination.
- In the case of a device that is not significant risk (NSR), the sponsor shall submit a final report to all reviewing IRBs within 6 months after termination or completion.

Unanticipated Adverse Events and Withdrawal of Approval

- The sponsor-investigator of an IDE must submit reports of unanticipated adverse device effects to the FDA and all reviewing IRBs.
- The sponsor-investigator of an IDE must submit reports regarding withdrawal of FDA approval to all reviewing IRBs.
- The sponsor-investigator of an IDE must submit reports regarding withdrawal of approval of any reviewing IRB to all reviewing IRBs and the FDA.
- What are the requirements for "emergency treatment use" of FDA test articles in clinical care situations without prior convened JHM IRB review?

The JHM IRB may allow for "emergency treatment use" of a test article with an IND or IDE in accordance with FDA regulations. For drugs, the situation must meet the definition of "Emergency Use" as stated in 21 CFR 56.102(d) :

"Emergency use means the use of a test article on a human subject in a life-threatening situation in which no standard acceptable treatment is available, and in which there is not sufficient time to obtain IRB approval."

For devices, in the case of serious disease, a device may ordinarily be made available for treatment use after all clinical trials are completed; in the case of an immediately life-threatening condition, the device may be made available before completion of clinical trials (21 CFR 812.36).

THE JHM IRB PREFERS NOTIFICATION OF AN IRB CHAIRPERSON PRIOR TO THE EMERGENCY USE WHEN TIME PERMITS

Emergency Use with Prior Notification or Approval by JHM IRB

Requests for emergency off-protocol use of FDA IND or IDE test articles outside of an IRB approved protocol must come from the JHH attending physician.

- The JHH attending physician must submit either a new eIRB application or if there is an existing eIRB application, a Further Study Action, Emergency Use request and must include a written statement the necessity for JHM IRB approval of an emergency use request.
- If the test article being used were not associated with a current research study, then the prescribing physician would need to submit a new eIRB application for the emergency use of the test article.
- If the test article is being used in a current study the prescribing physician must contact the PI of the current study and ask the PI to submit a Further Study Action activity in eIRB and advise the JHM IRB of the emergency use by submitting an Emergency Use Request.
- The new eIRB application for the emergency use or the Emergency Use Request would be reviewed by the JHM IRB at a convened meeting.
- The JHM IRB will not accept requests from non-attending physicians or staff.
- If time does not permit a JHM IRB convened review of the new eIRB application or Emergency Use Request, a JHM IRB Chairperson, Co-Chairperson and/or P & T representative may review the request for compliance with FDA regulations.
- If necessary, the reviewer will provide a written response to the investigator with notification whether the emergency use is "cleared" to proceed. The response could be in the form of an e-mail to the PI.
- If the use does not meet regulatory requirements, the reviewer will inform the PI that the emergency use may not take place.

- The eIRB application will be reviewed by the JHM IRB at a convened meeting whether or not the reviewer "clears" the use of the test article.
- If the PI receives clearance from the JHM IRB, the response will include a reminder that she/he must submit safety reports or adverse event reports to either the sponsor who holds the IND/IDE for the test article, or if the PI holds the IND/IDE, to the FDA as required by regulation. A copy of such reports should be uploaded into the eIRB application.

Emergency Use Without Prior Notification to or Approval by JHM IRB

Although the JHM IRB prefers to be notified before a test article is used in an emergency clinical care situation, there may be cases for which there is no time to notify the JHM IRB.

In such cases, the physician may proceed with the emergency use in accord with 21 CFR 56.104(c) :

- The prescribing physician must notify the JHM IRB of the emergency use within 5 working days after use of the test article.
- If the test article being used were not associated with a current research study, then the prescribing physician would need to submit a new eIRB application for the emergency use of the test article.
- If the test article is being used in a current study the prescribing physician must contact the PI of the current study and ask the PI to submit a Further Study Action activity in eIRB and advise the JHM IRB of the emergency use by submitting an Emergency Use Request.
- The new eIRB application for the emergency use or the Emergency Use Request would be reviewed by the JHM IRB at a convened meeting.
- Consent for emergency use of a test article must be obtained from the subject or their legally authorised representative (LAR) using a research consent form. A Johns Hopkins Hospital consent form may not be used to obtain such consent.
- A research consent document provided either by a sponsor or by the physician must contain all of the FDA required elements.
- The research consent form must be reviewed and signed by the subject or subject's legally authorised representative even in emergency use cases.
- In the rare case where the test article is administered or used prior to consent, the treating physician must still obtain a research consent post administration or use.
- The JHM IRB will provide a written response acknowledging the physician's submission of the emergency request information. The acknowledgement will remind the physician that she/he must submit safety reports or adverse event reports to either the sponsor who holds the IND/IDE for the test article, or if the PI holds the emergency IND, to the FDA

as required by regulation. A copy of such reports should be uploaded into the eIRB application.

What are the Limits on Use of Data Obtained from Emergency Use of a Test Article Under the FDA Exemption

To be exempt from the FDA requirement (21 CFR 56.104) for IRB review for the emergency use of a test article in a life-threatening situation, an investigator must not use the data in a systematic investigation designed to develop or contribute to generalisable knowledge or else the exemption no longer applies.

To comply with this limitation, investigators must follow these rules :

- If the emergency use of the test article occurred without prior IRB notification and approval, the prescribing physician must notify the JHM IRB of the emergency use within 5 working days after use of the test article;
- Do not use the FDA emergency use exemption from IRB review to circumvent the general requirement for prior IRB review;
- Do not use data from an emergency use of a test article in a prospective research study; and
- Do not report data from an emergency use of a test article in a retrospective research study, unless the PI obtains specific approval from the IRB.
- What are the requirements for "compassionate use" of FDA test articles in clinical care situations?

Compassionate use" is not an Official FDA term.

Compassionate Use Generally Implies the Use of Non-FDA Approved Test Articles for a Single Patient or a Small Group of Seriously Ill Patients, Where no Other Available Treatments are Satisfactory

1. *A compassionate use of a test article may be needed if* :
    - There are no other FDA approved test articles available for the disease or condition;
    - The patient does not meet the inclusion criteria of a study using an IDE; or
    - There are no alternative treatments.

IRB and FDA Review and Approval of Compassionate Use of Test Articles must be Obtained Before the Test Article Can be Used

The PI may use a non-FDA approved, compassionate use test article in order to treat, diagnose, or monitor a patient with a serious disease or condition. However, the PI must justify that the probable risk of using the compassionate use test article to the patient is not greater than the probable risk from the disease.

The PI will be required to submit an eIRB application or a Further Study Action, Change in Research activity, and request the compassionate use of a test article.

The PI would have to upload the following documents :

- Sponsor approval;
- FDA approval;
- Independent assessment from an independent attending physician describing the need for the compassionate use of the test article.

Consent for compassionate use of a test article must be obtained from the subject or their authorised representative using a research consent form. Consent for compassionate use of a humanitarian use device or HUD must be obtained from the participant or LAR using the standard hospital consent form, and does not require the use of a research consent form.

WHAT ARE THE REQUIREMENTS FOR "HUMANITARIAN USE DEVICE" IN CLINICAL CARE SITUATIONS

A humanitarian use device (HUD) is commercially approved under the Humanitarian Use Exemption (HDE).

The HUD is not considered "investigational" nor is the use of the device considered "research." However, the federal regulations require IRB review and approval before a HUD is used.

The PI is required to submit an eIRB application in order for the IRB to review and approve the use a HUD :

- The IRB does not need to review and approve each individual use of a HUD, as long as the HUD is used within the FDA approved indication.
- However, if the IRB desires, it may require approval of the device on a case-by-case basis.

If the HUD is to be used in an emergency situation and the PI does not have time to obtain IRB approval before the use of the HUD, the PI is required to submit an eIRB application for IRB review within 5 days after the emergent use of the HUD. The PI must include an explanation as to why prior IRB approval was not obtained.

Consent for the use of a HUD (with prior IRB approval or in an emergency situation) must be obtained from the participant or LAR using the standard hospital consent form, and does not require the use of a research consent form.

THE FDA'S DRUG REVIEW PROCESS : ENSURING DRUGS ARE SAFE AND EFFECTIVE

The path a drug travels from a lab to your medicine cabinet is usually long, and every drug takes a unique route. Often, a drug is developed to treat a specific disease. An important use of a drug may also be discovered by accident.

For example, Retrovir (zidovudine, also known as AZT) was first studied as an anti-cancer drug in the 1960s with disappointing results. It wasn't until the 1980s that researchers discovered the drug could treat AIDS, and the Food and Drug Administration approved the drug, manufactured by Glaxo Smith Kline, for that purpose in 1987.

Most drugs that undergo pre-clinical (animal) testing never even make it to human testing and review by the FDA. The drugs that do must undergo the agency's rigourous evaluation process, which scrutinises everything about the drug—from the design of clinical trials to the severity of side effects to the conditions under which the drug is manufactured.

Stages of Drug Development and Review

Investigational New Drug Application (IND)—The pharmaceutical industry sometimes seeks advice from the FDA prior to submission of an IND. Sponsors—companies, research institutions, and other organisations that take responsibility for developing a drug—must show the FDA results of preclinical testing they've done in laboratory animals and what they propose to do for human testing. At this stage, the FDA decides whether it is reasonably safe for the company to move forward with testing the drug in humans.

Clinical Trials

Drug studies in humans can begin only after an IND is reviewed by the FDA and a local institutional review board (IRB). The board is a panel of scientists and non-scientists in hospitals and research institutions that oversees clinical research.

IRBs approve the clinical trial protocols, which describe the type of people who may participate in the clinical trial, the schedule of tests and procedures, the medications and dosages to be studied, the length of the study, the study's objectives, and other details. IRBs make sure the study is acceptable, that participants have given consent and are fully informed of their risks, and that researchers take appropriate steps to protect patients from harm.

Phase 1 studies are usually conducted in healthy volunteers. The goal here is to determine what the drug's most frequent side effects are and, often, how the drug is metabolised and excreted. The number of subjects typically ranges from 20 to 80.

Phase 2 studies begin if Phase 1 studies don't reveal unacceptable toxicity. While the emphasis in Phase 1 is on safety, the emphasis in Phase 2 is on effectiveness. This phase aims to obtain preliminary data on whether the drug works in people who have a certain disease or condition. For controlled trials, patients receiving the drug are compared with similar patients receiving a different treatment—usually an inactive substance (placebo), or a different drug. Safety continues to be evaluated, and short-term side effects are studied. Typically, the number of subjects in Phase 2 studies ranges from a few dozen to about 300.

At the end of Phase 2, the FDA and sponsors try to come to an agreement on how the large-scale studies in Phase 3 should be done. How often the FDA meets with a sponsor varies, but this is one of two most common meeting points prior to submission of a new drug application. The other most common time is pre-NDA—right before a new drug application is submitted. Phase 3 studies begin if evidence of effectiveness is shown in Phase 2. These studies gather more information about safety and effectiveness, studying different populations and different dosages and using the drug in combination with other drugs. The number of subjects usually ranges from several hundred to about 3,000 people.

Post-market requirement and commitment studies are studies required of or agreed to by a sponsor that are conducted after the FDA has approved a product for marketing. The FDA uses post-market requirement and commitment studies to gather additional information about a product's safety, efficacy, or optimal use.

New Drug Application (NDA)

This is the formal step a drug sponsor takes to ask that the FDA consider approving a new drug for marketing in the United States. An NDA includes all animal and human data and analyses of the data, as well as information about how the drug behaves in the body and how it is manufactured.

When an NDA comes in, the FDA has 60 days to decide whether to file it so that it can be reviewed. The FDA can refuse to file an application that is incomplete. For example, some required studies may be missing. In accordance with the Prescription Drug User Fee Act (PDUFA), the FDA's Center for Drug Evaluation and Research (CDER) expects to review and act on at least 90 per cent of NDAs for standard drugs no later than 10 months after the applications are received. The review goal is six months for priority drugs.

"It's the clinical trials that take so long—usually several years," says Sandra Kweder, M.D., deputy director of the Office of New Drugs in the CDER. "The emphasis on speed for FDA mostly relates to review time and timelines of being able to meet with sponsors during a drug's development," she says.

Reviewing Applications

Though FDA reviewers are involved with a drug's development throughout the IND stage, the official review time is the length of time it takes to review a new drug application and issue an action letter, an official statement informing a drug sponsor of the agency's decision.

Once a new drug application is filed, an FDA review team—medical doctors, chemists, statisticians, microbiologists, pharmacologists, and other experts—evaluates whether the studies the sponsor submitted show that the drug is safe and effective for its proposed use. No drug is absolutely safe; all drugs have side effects. "Safe" in this sense means that the benefits of the drug appear to outweigh the risks.

The review team analyses study results and looks for possible issues with the application, such as weaknesses of the study design or analyses. Reviewers determine whether they agree with the sponsor's results and conclusions, or whether they need any additional information to make a decision.

Each reviewer prepares a written evaluation containing conclusions and recommendations about the application. These evaluations are then considered by team leaders, division directors, and office directors, depending on the type of application.

Reviewers receive training that fosters consistency in drug reviews, and good review practices remain a high priority for the agency.

Sometimes, the FDA calls on advisory committees, who provide FDA with independent opinions and recommendations from outside experts on applications to market new drugs, and on FDA policies. Whether an advisory committee is needed depends on many things.

"Some considerations would be if it's a drug that has significant questions, if it's the first in its class, or the first for a given indication," says Mark Goldberger, M.D., a former director of one of CDER's drug review offices. "Generally, FDA takes the advice of advisory committees, but not always," he says. "Their role is just that—to advise."

Accelerated Approval

Traditional approval requires that clinical benefit be shown before approval can be granted. Accelerated approval is given to some new drugs for serious and life-threatening illnesses that lack satisfactory treatments. This allows an NDA to be approved before measures of effectiveness that would usually be required for approval are available.

Instead, less traditional measures called surrogate endpoints are used to evaluate effectiveness. These are laboratory findings or signs that may not be a direct measurement of how a patient feels, functions, or survives, but are considered likely to predict benefit. For example, a surrogate endpoint could be the lowering of HIV blood levels for short periods of time with anti-retroviral drugs.

Gleevec (imatinib mesylate), an oral treatment for patients with a life-threatening form of cancer called chronic myeloid leukemia (CML), received accelerated approval. The drug was also approved under the FDA's orphan drug programme, which gives financial incentives to sponsors for manufacturing drugs that treat rare diseases. Gleevec blocks enzymes that play a role in cancer growth. The approval was based on results of three large Phase 2 studies, which showed the drug could substantially reduce the level of cancerous cells in the bone marrow and blood.

Most drugs to treat HIV have been approved under accelerated approval provisions, with the company required to continue its studies after the drug is on the market to confirm that its effects on virus levels are maintained and that it ultimately benefits the patient. Under accelerated approval rules, if studies don't confirm the initial results, the FDA can withdraw the approval.

Because pre-market review can't catch all potential problems with a drug, the FDA continues to track approved drugs for adverse events through a postmarketing surveilla nce programme.

Bumps in the Road

If the FDA decides that the benefits of a drug outweigh the risks, the drug will receive approval and can be marketed in the United States. But if there are problems with an NDA or if more information is necessary to make that determination, the FDA may issue a complete response letter.

Common problems include unexpected safety issues that crop up or failure to demonstrate a drug's effectiveness. A sponsor may need to conduct additional studies—perhaps studies of more people, different types of people, or for a longer period of time.

Common problems include unexpected safety issues that crop up or failure to demonstrate a drug's effectiveness. A sponsor may need to conduct additional studies—perhaps studies of more people, different types of people, or for a longer period of time.

Manufacturing issues are also among the reasons that approval may be delayed or denied. Drugs must be manufactured in accordance with standards called good manufacturing practices, and the FDA inspects manufacturing facilities before a drug can be approved. If a facility isn't ready for inspection, approval can be delayed. Any manufacturing deficiencies found would need to be corrected before approval.

"Sometimes a company may make a certain amount of a drug for clinical trials. Then when they go to scale up, they may lose a supplier or end up with quality control issues that result in a product of different chemistry," says the FDA's Kweder. "Sponsors have to show us that the product that's going to be marketed is the same product that they tested."

John Jenkins, M.D., director of CDER's Office of New Drugs, says, "It's often a combination of problems that prevent approval." Close communication with the FDA early on in a drug's development reduces the chance that an application will have to go through more than one cycle of review, he says. "But it's no guarantee."

The FDA outlines the justification for its decision in a complete response letter to the drug sponsor and CDER gives the sponsor a chance to meet with agency officials to discuss the deficiencies. At that point, the sponsor can choose to ask for a hearing, or correct any deficiencies and submit new information, or they can withdraw the application.

DRUG REVIEW STEPS

Pre-clinical (Animal) Testing

An investigational new drug application (IND) outlines what the sponsor of a new drug proposes for human testing in clinical trials :

- Phase 1 studies (typically involve 20 to 80 people).
- Phase 2 studies (typically involve a few dozen to about 300 people).
- Phase 3 studies (typically involve several hundred to about 3,000 people).
- The pre-NDA period, just before a new drug application (NDA) is submitted. A common time for the FDA and drug sponsors to meet.
- Submission of an NDA is the formal step asking the FDA to consider a drug for marketing approval.
- After an NDA is received, the FDA has 60 days to decide whether to file it so it can be reviewed.
- If the FDA files the NDA, an FDA review team is assigned to evaluate the sponsor's research on the drug's safety and effectiveness.
- The FDA reviews information that goes on a drug's professional labeling (information on how to use the drug).
- The FDA inspects the facilities where the drug will be manufactured as part of the approval process.
- FDA reviewers will approve the application or issue a complete response letter.

The Role of User Fees

Since the Prescription Drug User Fee Act (PDUFA) was passed in 1992, more than 1,000 drugs and biologics have come to the market, including new medicines to treat cancer, AIDS, cardiovascular disease, and life-threatening infections. PDUFA has allowed the Food and Drug Administration to bring access to new drugs as fast or faster than anywhere in the world, all while maintaining the same thorough review process.

Under PDUFA, drug companies agree to pay fees that boost FDA resources, and the FDA agrees to time goals for its review of new drug applications. Along with supporting increased staff, drug user fees help the FDA upgrade resources in information technology. The agency has moved towards an electronic submission and review environment, now accepting more electronic applications and archiving review documents electronically.

The goals set by PDUFA apply to the review of original new human drug and biological applications, re-submissions of original applications, and supplements to approved applications. The second phase of PDUFA, known as PDUFA II, was re-authorised in 1997 and extended the user fee programme through September 2002. PDUFA III, which extended to Sept. 30, 2007, was reauthorised in June 2002.

PDUFA III allowed the FDA to spend some user fees to increase surveillance of the safety of medicines during their first two years on the market, or three years for potentially dangerous medications. It is during this initial period, when new medicines enter into wide use, that the agency is best able to identify and counter adverse side effects that did not appear during the clinical trials.

On September 27, 2007, President Bush signed into law the Food and Drug Administration Amendments Act of 2007 which includes the re-authorisation and expansion of the Prescription Drug User Fee Act. The re-authorisation of PDUFA will significantly broaden and upgrade the agency's drug safety programme, and facilitate more efficient development of safe and effective new medications for the American public.

In addition to setting time frames for review of applications, PDUFA sets goals to improve communication and sets goals for specific kinds of meetings between the FDA and drug sponsors. It also outlines how fast the FDA must respond to requests from sponsors. Throughout a drug's development, the FDA advises sponsors on how to study certain classes of drugs, how to submit data, what kind of data are needed, and how clinical trials should be designed.

The Quality of Clinical Data

The Food and Drug Administration relies on data that sponsors submit to decide whether a drug should be approved. To protect the rights and welfare of people in clinical trials, and to verify the quality and integrity of data submitted, the FDA's Division of Scientific Investigations (DSI) conducts inspections of clinical investigators' study sites. DSI also reviews the records of institutional review boards to be sure they are fulfilling their role in patient protection.

"FDA investigators compare information that clinical investigators provided to sponsors on case report forms with information in source documents such as medical records and lab results," says Carolyn Hommel, a consumer safety officer in DSI.

DSI seeks to determine such things as whether the study was conducted according to the investigational plan, whether all adverse events were recorded, and whether the subjects met the inclusion/exclusion criteria outlined in the study protocol.

At the conclusion of each inspection, FDA investigators prepare a report summarising any deficiencies. In cases where they observe numerous or serious deviations, such as falsification of data, DSI classifies the inspection as "official action indicated" and sends a warning letter or Notice of Initiation of Disqualification Proceedings and Opportunity to Explain (NIDPOE) to the clinical inve stigator, specifying the deviations that were found.

The NIDPOE begins an administrative process to determine whether the clinical investigator should remain eligible to receive investigational products and conduct clinical studies.

CDER conducts about 300–400 clinical investigator inspections annually. About 3 per cent are classified in this "official action indicated" category.

The FDA has established an independent Drug Safety Oversight Board (DSOB) to oversee the management of drug safety issues. The Board meets monthly and has representatives from three FDA Centers and five other federal government agencies. The board's responsibilities include conducting timely and comprehensive evaluations of emerging drug safety issues, and ensuring that experts—both inside and outside of the FDA—give their perspectives to the agency. The first meeting of the DSOB was held in June 2005.

Chapter 15

Determination of Soil Enzymes Activity

INTRODUCTION

Soil enzymes increase the reaction rate at which plant residues decompose and release plant available nutrients. The substance acted upon by a soil enzyme is called the substrate. For example, glucosidase (soil enzyme) cleaves glucose from glucoside (substrate), a compound common in plants. Enzymes are specific to a substrate and have active sites that bind with the substrate to form a temporary complex. The enzymatic reaction releases a product, which can be a nutrient contained in the substrate.

Sources of soil enzymes include living and dead microbes, plant roots and residues, and soil animals. Enzymes stabilised in the soil matrix accumulate or form complexes with organic matter (humus), clay, and humus-clay complexes, but are no longer associated with viable cells. It is thought that 40 to 60 per cent of enzyme activity can come from stabilised enzymes, so activity does not necessarily correlate highly with microbial biomass or respiration. Therefore, enzyme activity is the cumulative effect of long-term microbial activity and activity of the viable population at sampling.

However, an example of an enzyme that only reflects activity of viable cells is dehydrogenase, which in theory can only occur in viable cells and not in stabilised soil complexes.

MEASURING SOIL ENZYMES

Enzymes are measured indirectly by determining their activity in the laboratory using biochemical assays. Enzyme assays reflect potential activity and do not represent true *in situ* activity levels and must be viewed as an index.

When possible, compare the site of interest to samples taken from an adjacent, undisturbed site on the same soil type. Alternatively, for a newly implemented land management system, track changes from time zero to five or more years with annual sampling to detect temporal changes in activity of soil enzymes.

SOIL QUALITY

Soil quality is conditioned by the properties of the soil and the kind of climate and micro-climate. All soil properties, such as macro- and micro-morphological, physical, chemical, physico-chemical, biological, microbiological and biochemical, play an important part in the formation of soil quality. These properties are related to each other and are affected by the kind of parent rock and the specific climate conditions. In trying to determine soil quality, taking into account the properties which come into play here as integral parts is impossible. Biological, microbiological and biochemical properties depend on physical, chemical and physicochemical soil properties and the type of climate and micro-climate of the environment. Besides, Trasar-Cepeda *et. al.* stated that soil biological and biochemical properties are highly sensitive to environmental stress and thus can be used to assess soil quality.

Hitherto there have been few attempts to develop soil quality indices based on biological properties. In the majority of cases these assessments were made based on the activity of tillage soil horizons. Arable soils differ from forest soils in the morphology of a soil profile and, as a consequence, soil properties. Arable soils leave a relatively homogeneous, mineral-humus tillage horizon, reaching usually 30 cm deep. Forest soils, according to their biological activity and subsequent rate of turnover of organic matter entering to soil, leave or do not leave surface organic layers. The comparison of soil properties in differentiated soil profiles on the basis of surface layers is groundless and leads to incorrect conclusions.

Usually the highest microbiological activity in soils takes place in the surface horizon. Studies carried out by one author showed that the highest enzyme activities in some subtypes of forest soils in deeper horizons are noted and endopedons lying to 1 m deep leave appreciable enzyme activity and can have a serious impact on the rate of transformation in soils and, consequently, on soil quality. For that reason, in these studies the activity of enzymes in natural moisture soil samples from particular horizons of profiles were determined. Then enzyme activities were counted on the basis of oven-dry weight soil mass volume, in a column at a area of 1 dm^2, and to 1 m deep.

The authors of papers which evaluate soil quality based on of enzyme activity agree that determination of a single soil enzyme could not be used to determine plant productivity. From own investigations on enzyme activities in mor, moder and mull humus horizons of forest soils of the Western Beskid Mts. It appears that activity of dehydrogenases, urease and proteases in account in the soil volume, reflected correctly the biological activity of the investigated types of humus. It was found that dehydrogenase, urease and protease activities expressed in terms of soil volume in a column of $1dm^2$ - 1 m ($10\ dm^3$), fully reflected the quality of forest soils of southern Poland. To estimate quality of the soils, the following enzyme soil index was proposed :

$$SEI_{DPU} = 0.01 \times DhA + PA + UA,$$

Where : DhA - Dehydrogenase activity in mg TPF $10\ dm^{-3}$ 24 hs, PA - Protease activity in g $N\text{-}NH_2$ $10\ dm^{-3}$ $20\ hs^{-1}$, and UA - Urease activity in g $N\text{-}NH_4$ $10\ dm^{-3}$ $24\ hs^{-1}$.

The Index Tested Some Sub-types of Podzol and Brown Soil, and Gave Reliable Evaluation of Its Quality

The aim of the study was a verification of the usefulness of the SEI_{DPU} to assessment of soil quality on the base of other forest soil subtypes of southern Poland, extremely differentiated according to their quality. The study included the following soil types : podzol, lessivés, brown, rendzin and extremely fertile humus-deluvial soil, found in a complex of arable chernozem, which is unusual in forest areas. The aim of the study was also to investigate the effects of forest stand composition on soil enzyme activity and soil quality. Therefore, enzyme activities of podzols, lessivés and acid brown soils under conifers and deciduous forest stand were investigated.

AREA DESCRIPTION, METHODS AND MATERIAL STUDIED

The investigations were carried out in the summer periods of the years 1997–2001. The studies on the same sub-types of soil were conducted on the same day of the same year. The study included :

- Two proper podzol soils (*Haplic Podzols*), the first one under the monoculture of pine forest stand (plot No. 1) and the second under the pine forest stand with an admixture of birch and beech understory (plot No. 2) in forest compartment No. : 58c and 60a respectively, of Rudy Raciborskie Forest District;
- Two lessivés soils, the first podzolised lessivés soil (*Hyperdistric Luvisol*) under a larch forest stand (plot No. 3) and the second browned lessivés soil (*Epidystric Luvisol*) under the forest stand of oak and hornbeam as the climax vegetation, location in the forest compartment 57l and 51i, respectively of Miechów Forest District;
- Two acid brown soils (*Hyperdistric Cambisols*), the first under a monoculture spruce forest stand (plot No. 4) and the second under the beech forest stand (plot No. 5), located in the forest compartments 70f and 70g, respectively of Ustroñ Forest District;
- Two leached brown soils (*Cambisols*), the first with the fir-beech forest stand as the climax vegetation, located in the area of control unit No. 121 of Forest Didactic Station Krynica - *Epidystrict Cambisol* and the second

under the lime forest stand as a climax vegetation located in the area of "Obro¿yska" Reservation of Piwniczna Forest District—
Endoeutric Cambisol ;

- Chalk rendzina (*Rendzic Leptosol*) under a larch-hornbeam forest stand located in forest compar tment 18a of Krasnystaw Forest District,
- Humous deluvial soil (*Hypereutric Luvisol*) under ash forest stand located in forest compartment 58c of Miechów Forest District.

For every plot, a soil profile has been done, which have received the same numbers as their plots. In all profiles the soil horizons were marked according to methods used in soil science. The contents of the soil skeleton (gravel and stone) in field conditions, in volume per cent and with reference to total soil mass, were estimated visually. Soil samples from every horizon were collected in plastic bags, and samples of natural structure, to cores of 250 cm^3 volume, in order to determine the soil bulk density were collected as well. In soil horizons with plenty (> 60 per cent) of soil skeleton, soil without skeleton was collected in its field condition to measuring cylinders, imitating the natural density by compression.

Soil samples from surface horizons were collected from nine places around the soil profiles and then homogenised. The samples from other horizons were collected inside soil profiles. The samples were transported to the laboratory on the same day and stored in a refrigerator at 4°C. In a laboratory, one part of these samples was sieved through a 2-mm sieve. *In samples with a natural moisture content, before no more then a few days, the activity of the following enzymes was colourimetrically determined* :

- Dehydrogenases by the Lenhard method using the Casida *et. al.* procedure;
- Proteases by the Hoffmann and Teicher method;
- Urease by the Romeiko and Malinska method with Èunderova modification, using phosphate buffer at pH 6.7, 5 per cent urea solution as substrate and 24-hour incubation.

All incubations were carried out at 37°C. The moisture content was determined by weight method. Enzyme activities were expressed in terms of oven dry soil volume in a column of 1dm^2 and natural soil depth to a maximum of 1 m of deep. In these calculations the volume of soil skeleton was deducted and oven dry soil mass was calculated according to soil bulk density.

The second part of collected samples was dried at room temperature and then the samples were sieved through a 2-mm sieve. In these samples physical, physicochemical and chemical properties necessary to the undertaking of typological diagnoses were determined.

To determine of these properties the following methods were used :

- Texture using the Bouyocous-Casagrande method, as modified by Proszynski method;
- Soil pH in H_2O and 1M KCl solution with soil to solution ratio in organic soil as 1 : 5, in sand 1 : 1, and in loam soil as 1 : 2.5 by potentiometric method;

- Cation exchange capacity (CEC) was calculated as a sum of the total acidity by Kappen method and sum of base cations (BC) in 1M CH_3COONH_4 at pH 7.0 as extractor and degree of base saturation (BS per cent = BC/CEC × 100) was calculated;
- Organic C by the Tiurin method with Soil Science Department of Agriculture University modification (without use of catalyser and longer time of oxidation),
- Total nitrogen by the Kjeldahl method.

DISCUSSION AND CONCLUSIONS

The comparison of enzyme activities which was investigated in the same summer period but in different years can be questionable. The results of earlier conducted investigations revealed a minor difference in the activity of the investigated enzymes between the analysed years.

The results of the investigations of Gianfreda *et. al.* and the revue of Kiss *et. al.* pointed out the individual sorption abilities of enzymes to different minerals and the different residual activities of enzymes after accumulation. Zantua and Bremner, quotation by Bremner and Mulvaney stated, "urease activity increased after addition of jack bean urease or of organic materials promoting microbial activity, but subsequently decreased and stabilised at the level observed before addition of these substances" and the explanation of these findings is that "the urease activity of unamended soils reflects their capacity for protection of urease and that urease in excess of this capacity is decomposed or inactivated" [3]. According to these results, can be hypothesised that soil sub-types can heave a peculiar capacity towards individual enzyme dependence on some kinds of minerals and may be kinds of humus colloids.

The microbial activity of soil, to a large extent, is dependent on the quantity of available carbon, and is reflected by dehydrogenase activity. Nitrogen mineralisation rates have been found to be highly correlated with plant nitrogen uptake and net aboveground productivity - Pastor *et. al.* 1984, quotation of White. The potential rate of nitrogen mineralisation in soil is reflected by the potential activities of protease and urease. Consequently, the evaluation of soil quality based on dehydrogenase, protease and urease activities seem to be well justified.

A broad range of received values of SEI_{DPU} gives the possibility of elaborating on biochemical soil classification. In the Table quality soil classes are proposed, conditioned by the range of SEI_{DPU} values. Because on forest grounds soils of low quality dominate, narrow class sections (6 units) are proposed, allowing the possibility to classify poorer forest soil, and to a lesser degree proposed classes can divide rich soil, which is rarely found on forest grounds, but form rather arable ecosystems. Therefore the proposed division is connected exclusively to the forest soil of southern Poland.

Table. Enzyme Soil Index, SEI_{DPU}, vs. Quality of Some Investigated Forest Soil Subunits of Southern Poland.

SEI_{DPU}	Soil subunits	Profile No.	Soil quality
< 6	Haplic podzol, Hyperdistric Luvisol, Hyperdistric Cambisol	1–4	Very low
6–12	Hyperdistric Cambisol	5	Low
12–18	Epidystric Cambisol, Endoeutric Cambisol	7, 8	Medium
18–24	Epidystric Luvisol	6	High
> 24	Rendzic Leptosol, Hypereutric Luvisol	9, 10	Very high

The sequence of investigated soil sub-types in increasing order of SEI_{DPU} values, deviating to a small degree from the sequence of sub-types in Table is ordered according to the commonly acknowledged quality of these soils. The Hyperdistric Luvisol under the larch forest stand, received relatively low values of SEI_{DPU} on the level of Haplic Podzol under pine monoculture. A appreciable decrease of enzyme activities can be observed in soil under monoculture conifer trees in the pH of soil, in 1M KCl below 3.5. These low values of SEI_{DPU} for podzolised lessivés soil can be explained by the influence of slow decomposing larch needles.

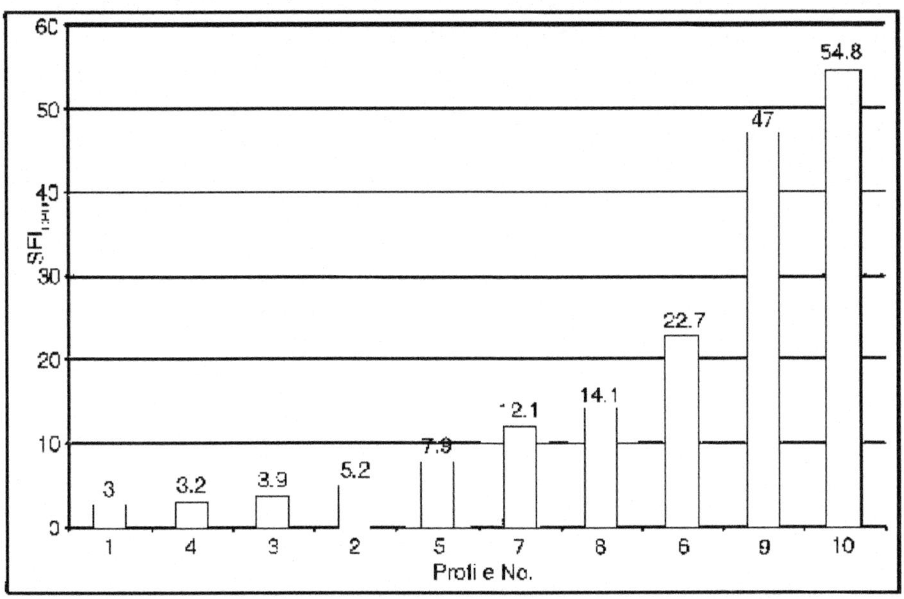

Fig. Soil Enzyme Index (SEI_{DPU}) Established Based on Dehydrogenase, Protease and Urease Activities for Some Forest Soil Subunits of Southern Poland.

Larch litter has the highest C : N ratio (C : N = 77) among the species which make up forests in Europe. The unprofitable effect of the monoculture of conifer forest stands on the enzyme activity of soil can be observed as well as in the examples of Cambisols under the spruce monoculture and the beech forest stand (Profiles 4 and 5), Haplic Podzols under the pine monoculture and the mixed forest stand (Profiles 1 and 2). Chodzicki mentioned about the profitable effect of beech on soils properties in pine forests. The profitable interaction of birch leaves in comparison to pine needles on nutrients oven time, especially of nitrogen and phosphorus by Berg and Staaf as well as by Ste-Marie and Paré have been described. Steubing has been written on many substances which contain pine needles, in comparison to beech leaves, which retarded microbiological attacks.

Epidystric Luvisol (browned lessivés soil) received relatively high SEIDPU values under the climax deciduous forest stand. The lessivé soils are known for their rapid rate of mineralisation of soil organic matter and low content of humus, very often accumulated in this soil in a kind of ochric diagnostic horizon. An intensive rate of mineralisation of organic matter in lessivés soils can be caused by a higher moisture content in surface horizons, due to a lesser permeability of argic horizons.

The highest value of SEIDPU, as expected, was noted in Hypereutric Luvisol, deep humus deluvial soil (Profile 10). In soil with neutral and weak alkaline reactions there were, very advantageous physical properties, without any signs of gley process, while susceptible to the microbiological decomposition of leaves ash (C : N = 24).

Doubts as appeared with useful biochemical properties to evaluation of soil quality, as recently have been presented by some researchers, can result from evaluation of soil quality based on biochemical properties of humus horizons only, presented on weight but not volume units and leave out in evaluation other horizons in soil profile.

Soil can be seen as a perfect reactionary enzymatic column, naturally restored in enzyme and substrates. The substrates input to soil have a greater possibility to be incorporated into the accumulated enzymes in soils which are more and deeper enzymatic active. Therefore soil depth, enzyme active, may have a crucial role in the rate of enzyme transformation in soil. This is the reason for the assumption of soil quality, based on biochemical properties, the greater part of the soil profile of not only the humus horizon, which in forest soils is strongly differentiated according to quality and width.

From the studies curried out, it appears that an extremely critical matter is the adaptation of a suitable species composition for a given forest site, according to soil quality. Unsuitable species compositions of forest stands (excessive concentration of conifer species) in fertile soil can cause soil degradation during a period of one generation. The first phase sees a degradation of biological and biochemical soil properties, then physico-chemical, chemical and physical properties, and in the next phase micro-morphological and the end sees a change of morphological properties, which are easily visible.

For these reasons, to gain a true sense of reliable assessment of biological and biochemical properties of soil, the aim is fast detection of degradation, so that harmful interaction can be detected in the first stages of development.

Based on the investigations carried out, may form following conclusions :

- The proposed soil enzyme index (SEI_{DPU}), established based on dehydrogenase, protease and urease activities, gave reliable evaluation of quality of ten forest soil sub-types of southern Poland, and an interpretation of the order of these soils is possible, according to our present knowledge;
- A broad range of received values of SEIDPU allows the possibility of elaboration of biochemical soil classification,
- The monocultures of conifer trees, in comparison to deciduous forest stands, had disadvantageous effects unprofitable to the biochemical activity of soil, leading to its degradation. In investigated soils a great fall in enzyme activity can be observed in soil with pH in 1M KCl below 3.5;
- The adaptation of a proper species composition of forest stand to forest site according to soil quality, has a great effect on the state of soil quality and prevention of soil degradation,
- The evaluation of soil quality based on proposed SEIDPU would be tested on soils less drained than elsewhere.

Chapter 16

Organic Certified Compost to Control Soil-borne Diseases

INTRODUCTION

Organic farmers use on-farm resources whenever possible to increase organic matter in soil and to manage soil fertility, through the use of compost, cover crops and crop rotations. Some composts have been found to be suppressive against several soil-borne pathogens in various cropping systems. The use of compost as a peat substitute to control root pathogens in Italy was first suggested in 1988. An increase of some diseases due to compost usage has also been demonstrated, since compost is a product that varies considerably in chemical, physical and biotic composition, and, consequently, also in ability to suppress soil-borne diseases. The level and reproducibility of suppressive properties of compost can be increased upon antagonist enrichment. In the present study, a certified organic compost has been analysed for its suppressive activity against soil-borne diseases and compared to a not certified compost and to peat.

Materials and Methods

The composting plant "ACEA Pinerolese S.p.A." located in Pinerolo, near Torino (Italy), was selected, according to previous trials, for producing high suppressive compost. One organic certified compost and one not certified compost produced by the composting plant in three different periods (autumn, spring and summer) were selected for experiments. Organic certified compost was produced according to EC Reg. 2092/91 from biowaste, green and yard wastes, while non-certified compost was produced from sludge obtained from urban waste water treatment, organic domestic waste and green waste.

Compost maturity was estimated using Wood's End Lab's "Solvita" Compost Maturity Test Kit, which estimates maturity based on respiration rate. Germination and plant growth bioassays were used to test the quality of composts, according to Warman, 1999. The germination index (Gi) was calculated according to the formula Gi = G/G0 x L/L0 x 100, where G0 and L0 are respectively the germination percentage and radicle growth of the 100 per cent H2O control.

For the plant growth assays, compost was mixed with sphagnum peat at 20, 40 and 100 per cent v/v. Sphagnum peat was also used as control medium and the potting mixes were sowed with 100 seeds of Lepidium sativum. The plant growth index (Pi) was calculated according to the formula Pi = (PSc/PSp) x 100, where PSc is the average dry weight of plants grown on compost-amended potting mixes, while PSp is the average dry weight of plants grown on peat. The global plant growth index (Pi) was the Pi average of the 20, 40 and 100 per cent mixes treatments.

A second run of trials was carried out in greenhouse to test the suppressiveness of composts. A randomised block design was used. All assays were repeated at least twice. Five pots of 1 l of volume were prepared for each treatment and 7 seeds of tomato were sown in each pot. Seven days before sowing, pots were inoculated with Phytophthora nicotianae, previously propagated in flasks on wheat plus hemp and added to the pots at 2 g/l of mycelium plus substrate. Other pots were also prepared and 5 seeds of zucchini (Cucurbita pepo) were sown in each. Seven days before sowing, pots were inoculated with Phytophthora capsici at 2 g/l dosage. A sphagnum peat amendment was used as control mean. The disease levels were related to that of the peat control in order to be able to compare suppression levels among the different trials, based on percentage of disease suppression.

A third run of trials was carried out on an open field of bell pepper (Capsicum annuum) strongly infested in the previous years by Phytophthora capsici. Compost was inoculated 7 day before rototilling with several biological control agents at 4g/l dosages : Trichoderma harzianum T-22 (Rootshield®, 108 CFU/g, Intrachem Bio Italia), Trichoderma viride TV1 (TV1®, 108 CFU/g, Agribiotec) and T. harzianum (strain ICC012) in mixture with T. viride (strain ICC080) (Remedier®, 108 CFU/g, Isagro). Pepper plants were transplanted 2 days after organic certified compost (enriched or not with Trichoderma spp.) was rototilled into soil at 2 kg/m2 dosage. After transplanting diseased plants affected by P. capsici were counted every 7-10 days. The effect of micro-organisms added to compost on percentage of disease suppression was assessed compared to control.

Analyses of variance (ANOVA) were carried out with the statistical programme SPSS 12.0 (SPSS Inc., Chicago, IL). Tukey's HSD test was applied when one-way ANOVA revealed significant differences ($P<0.05$).

SOIL-BORNE DISEASES

The fungi have many important functions, largely unrecognised, in the biosphere. Although many species are beneficial, a considerable number are detrimental to our interests. There are approximately 100,000 species of fungi described in the literature and there are many, many more yet to be described. Many of these live saprophytically on dead organic matter on or in soil where they are regarded as the most important decomposers of plant residues and other organic matter. Many species produce the enzymes needed to degrade the lignin and cellulose in plant residues and so initiate the decomposition of these complex compounds.

More than 8,000 species of fungi are known to cause diseases of plants and most plants are susceptible to some fungal pathogens.

Some species of fungi, the mycorrhizae, live symbiotically on or in the roots of many plants. This relationship is basically parasitic but in many situations is probably beneficial to both the plant and the fungus. The growth of the plant is promoted by the improved uptake of some mineral nutrients while the fungus gains access to organic nutrients and shelter.

Fungi are common in soil, in air (mainly as spores) and on plant surfaces throughout the world in arid, tropical, temperate and alpine regions.

The diseases that are caused by fungal pathogens which persist (survive) in the soil matrix and in residues on the soil surface are defined as soil-borne diseases. Thus the soil is a reservoir of inoculum of these pathogens, the majority of which are widely distributed in agricultural soils. However, some species show localised distribution patterns. Damage to root and crown tissues is hidden in the soil. Thus these diseases may not be noticed until the above-ground (foliar) parts of the plant are affected severely showing symptoms such as stunting, wilting, chlorosis and death.

These diseases are difficult to control because they are caused by pathogens which can survive for long periods in the absence of the normal crop host, often have a wide host range including weeds, chemical control often does not work well, is not practical or is too expensive and it is difficult to develop resistant varieties of plants. These diseases are often very difficult to diagnose accurately and the pathogens may be difficult to grow in culture and identify accurately.

Morphology

Fungi are thread-like, multi-cellular, non-motile micro-organisms which lack chlorophyll. The basic unit of a fungus is the microscopic thread-like hypha, which is essentially a tube with a rigid wall enclosing cytoplasm and nuclei. In the higher fungi the hypha is divided into cells by regular cross-walls or septa. A small pore in each septum allows interaction between the cytoplasm in adjacent cells. The so-called "lower fungi", which are now known as members of the Kingdom Chromista do not have septa and so the cytoplasm flows freely along the hyphae. Hyphae grow by elongation of the tip which is thin and plastic, unlike the rigid, relatively thick walls of the older part of the hypha. As the hypha elongates it branches laterally, a short distance from the tip. The branches also elongate and produce further branches, a process which results in a radial pattern of growth. The hyphal tips of some fungi are able to penetrate intact cell walls of both living and dead plant tissue. The ability of a fungus to elongate from the hyphal tip, to branch and to penetrate intact surfaces enables it to radiate in all directions over and through a substrate such as plant tissue, dead organic matter, or a laboratory growth medium. This process of growth over or through a substrate is called colonisation, and the hyphae collectively constitute a colony. A new colony develops from a spore, hyphal fragment or a multi-cellular vegetative structure called a sclerotium. The size of a colony is determined by the nature and extent of the substrate colonised by the fungus.

Physiology and Ecology

Because the fungi lack chlorophyll they need an external source of organic compounds from which to synthesise the various materials needed for growth, that is, they are heterotrophic. The fungi secrete enzymes which break down solid materials into soluble compounds for absorption through their outer walls. They vary greatly in their ability to utilise different types of substrates. Some species are obligate parasites and are only able to utilise nutrients from living host tissue. The majority are saprophytic and obtain nutrients only from dead organic material such as plant residues. In addition there are species which are able to grow saprophytically or parasitically. Many species live saprophytically on residues on or in soil and have a major role in their decomposition.

Temperature, water potential and pH have a critical influence on the growth and survival of fungi. The optimum temperatures for growth of the majority of fungi are in the range 25 to 28°C with the minimum and maximum temperatures being 5 to 10°C and 33 to 35°C respectively. Thus most fungi cannot grow at the body temperature of man, 37°C. Optimum water potentials for growth range from -0.01 to -1.0 MPa depending on the species. Growth ceases between -5.0 and -15.0 MPa. Many fungi including important plant pathogens are able to grow at potentials which cause permanent wilting of crop plants. Slightly acidic conditions (pH 6 to 6.5) are optimum for the growth of most fungi. Although fungal growth ceases at low water potentials many species are adapted to survive quite well under such conditions, either as resistant spores, sclerotia, other specialised structures or as dormant hyphae in residues. In general, cold dry conditions favour fungal survival while hot wet conditions do not.

Reproduction

Most fungi are able to reproduce asexually and/or sexually, forming spores which serve as units for survival and/or dispersal. The spore can be defined as a small discrete propagative unit containing the genome of the fungus and sufficient nutrient reserves to enable the establishment of a new colony. Spores vary in size, shape, septation, colour and wall thickness. They are produced directly on the hyphae, on specialised hyphal branches, or on multi-cellular fruiting structures which range in size from approx. 0.2 mm to over 600 mm in diameter. Although the fungi are defined as non-motile, some species of the Oomycetes (lower fungi) produce motile spores called zoospores. These spores have small flagella which enable them to swim in water films over short distances.

Light and temperature commonly affect the formation of spores and fruiting structures. Temperature, and to a lesser extent light, also have a significant effect on hyphal colour and growth. Thus if colonies grow and reproduce under fluctuating conditions of temperature and light they may develop annular zonations with respect to colour, hyphal density and intensity of spore formation. The primary function of some spores is to ensure the long-term survival of the species in the absence of a suitable substrate, or during periods when other environmental conditions are unfavourable for growth. These spores are thick walled and remain where they form, germinating when contacted by a new substrate.

The primary function of the majority of the spores is, however, to enable geographic dispersal by rain splash or wind. Spores formed in a mucilaginous matrix are usually dispersed by rain splash, in small water droplets over a short distance. Wind dispersed spores are either ejected from a fruiting structure into the airstream or are formed in dry powdery masses which are swept up by air currents. Wind dispersal may carry spores over long distances but most are deposited within 100 metres of their origin. Deposition is a result of impaction, gravity or rain wash. Because a very high proportion of spores are deposited on surfaces unsuitable for fungal growth, vast numbers of spores are produced from each colony to ensure the continuity of the species when one substrate is exhausted.

Spores have an extremely important role in fungal diseases of plants. Airborne spores enable rapid spread of the fungus from plant to plant or crop to crop. Thick-walled spores such as chlamydospores are able to survive in soil for long periods and enable the pathogen to survive in the absence of a susceptible host. Some pathogens which affect crops grown over large areas, such as the stem-rust fungus of wheat, may produce vast quantities of spores which can be carried in the atmosphere as invisible spore clouds over hundreds of kilometres.

In addition to resistant spores the fungi have evolved various other modes of survival. Some species form resting-bodies called sclerotia, which consist of aggregations of hyphal cells into dense, rounded or irregular structures, generally ranging in size from 0.1 mm to 20 mm. In other species fruiting structures such as perithecia act as survival structures and release spores when conditions become favourable for spore germination and growth.

Key Features of Soil-borne Plant Pathogens

Disease Symptoms

Many soil-borne fungal plant pathogens cause disease of the roots or stem disrupting the uptake and translocation of water and nutrients from the soil. Therefore they commonly cause similar symptoms to drought and nutrient deficiencies; these include wilting, yellowing, stunting and plant death. Thus symptoms such as this do not necessarily indicate that a pathogen is causing disease.

Host Range

The majority of these pathogens have a wide host range and may cause different types of diseases on different hosts. A particular species may cause severe root rot on one host but only cause superficial and symptomless infection on the roots of another host. Crop rotation is not necessarily an appropriate control measure for some of these pathogens because of their wide host range which may include weeds. However some pathogens have a very narrow host range, *e.g.* forma speciales of the wilt pathogen Fusarium oxysporum. However, many of these formae speciales can persist by symptomless colonisation of the cortex of roots of non-host plants.

Epidemiology and Survival

In general these fungi can persist for long periods in soil in the absence of susceptible crops. Some species persist mainly as resistant hyphae in plant residues, in large pieces of stubble or small fragments. The inoculum of such pathogens declines gradually as the residues decompose. Decomposition is hastened by warm wet conditions and mechanical disruption caused by cultivation. Many species also persist as propagules which are adapted for long-term survival in soil. These include thick walled spores such as oospores and chlamydospores, together with sclerotia and microsclerotia.

Environment of Diseases Caused by Soil-borne Plant Pathogens

Root Rot

These diseases are caused by a diverse group of fungi and related organisms. The most important genera include Pythium and Phytophthora, Rhizoctonia, Cylindrocladium and Armillaria. These diseases are characterised by a decay of the true root system; some pathogens are generally confined to the juvenile roots whilst others are capable of attacking older parts of the root system. Symptoms that are observable include wilting, leaf death and leaf fall, death of branches and limbs and in severe cases death of the whole plant. Some examples of these diseases are shown below.

Stem, Collar and Head Rots

These diseases are also caused by a diverse group of pathogens including species of Phytophthora, Sclerotium, Rhizoctonia, Sclerotinia, Fusarium and occasionally Aspergillus niger. The most obvious symptom of these diseases is the decay of the stem at ground level. Quite often this decay can lead to symptoms of wilting, death of leaves and to death of the plant. Some examples of these diseases are shown below. In tropical agricultural ecosystems these fungi can cause stem, foliage and head rots during warm, wet conditions. Phytophthora spp. for example, can cause diseases such as heart rot of pineapple, blight of potato and tomato and some fruit rots in these conditions. Similarly Rhizoctonia spp. can cause leaf blight in maise and head rot of cabbage in warm wet weather.

Wilt Diseases

The main species of fungi that cause these diseases are Fusarium oxysporum and Verticillium spp. The symptoms of these diseases include wilting of the foliage and internal necrosis of the vascular tissue in the stem of the plant. Some species of bacteria can also cause similar types of diseases. Some examples of these diseases are shown below.

Seedling Blights and Damping-off Diseases

Various common names are used for diseases of seedlings such as seedling blight and damping-off. The fungi that commonly cause seedling death include Pythium, Phytophthora, Rhizoctonia, Sclerotium rolfsii and less commonly Fusarium spp.. These fungi can infect the seedling during the germination, pre-emergence or post-emergence phases of seedling establishment. Environmental factors which inhibit germination and emergence usually increase disease severity. Thus cold conditions, dry, or very wet soils or a hard soil surface commonly lead to increased seedling disease. In northern Vietnam Pythium, Rhizoctonia and Sclerotium rolfsii are commonly associated with seedling death of vegetables such as beans, cabbages and other cruciferous crops, cucurbits and tomato.

SOIL-BORNE DISEASE MANAGEMENT IN ORGANIC VEGETABLE PRODUCTION

Soil-borne diseases can be a major limitation to crop production, particularly for vegetables. They are often difficult to control, even with conventional strategies. Fungal, plasmodiophorid, oomycete, and bacterial pathogens, as well as viruses and plant parasitic nematodes, may all cause soil-borne diseases. Important soil-borne fungal pathogens are species of Fusarium, Rhizoctonia, Verticillium, and Sclerotinia. Important Phytomyxea (plasmodiophorids) pathogens are species of Plasmodiophora and Spongospora. Important oomycete pathogens include Phytophthora and Pythium.

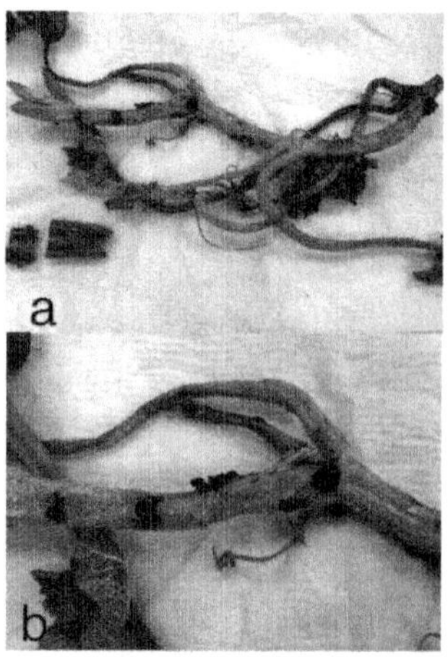

Fig. (a) Fusarium wilt of melon (Fusarium oxysporum). (b) Close-up of fusarium wilt on melon stem. Photo credit : Fulya Baysal-Gurel, The Ohio State University.

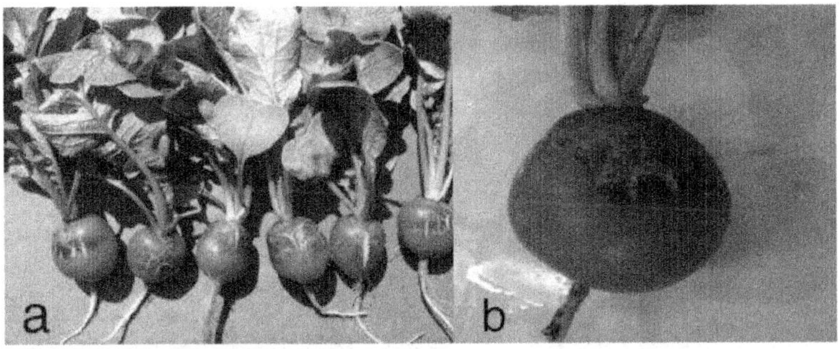

Fig. (a) Rhizoctonia root and hypocotyl rot of radishes (Rhizoctonia solani). (b) Close up of rhizoctonia root and hypocotyl rot on radish. Photo credit : Jhony Mera and Fulya Baysal-Gurel, The Ohio State University.

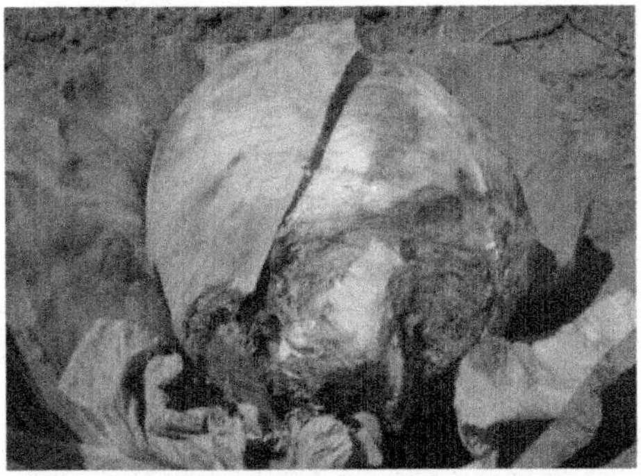

Fig. White mold of cabbage (Sclerotinia sclerotiorum). Photo credit : Jhony Mera, The Ohio State University.

Fig. Tomato white mold (Sclerotinia sclerotiorum). Photo credit : Fulya Baysal-Gurel, The Ohio State University.

Fig. (a) Lettuce drop (Sclerotinia minor). (b) Close up of lettuce drop (Sclerotinia minor) on lettuce. Photo credit : Sally A. Miller, The Ohio State University.

Fig. Clubroot of radish (Plasmodiophora brassicae). Photo credit : Jhony Mera, The Ohio State University.

Many soil-borne fungi persist in soil for long periods because they produce resistant survival structures such as melanised hyphae, chlamydospores, oospores, and sclerotia. Important soil-borne bacterial pathogens are Ralstonia, Pectobacterium, Agrobacterium, andStreptomyces. Pathogens in the Pseudomonas and Xanthomonas groups usually persist in the soil for only a short time. Soil-borne viruses that affect vegetables are few in number and generally survive only in the living tissues of the host plant or in the nematode or fungal vectors that transmit them. An exception is the Tombusvirus group, in which the viruses can survive in the soil without being inside a host plant or vector. Soil-borne plant parasitic nematodes affect vegetables by reducing plant vigour and growth. Important plant parasitic nematodes are root knot and cyst nematodes.

Fig. Phytophthora blight and fruit rot of pepper (Phytophthora capsici). Photo credit : Jhony Mera, The Ohio State University.

Fig. Phytophthora blight of winter squash (Phytophthora capsici). Photo credit : Jhony Mera, The Ohio State University.

Fig. Pepper damping-off (Pythium spp.). Photo credit : Fulya Baysal-Gurel, The Ohio State University.

Fig. Tomato bacterial wilt (Ralstonia solanacearum).
Photo credit : Sally A. Miller.

Soil-borne pathogens often survive for long periods on host plant debris, soil organic matter, or as free-living organisms. Each vegetable crop may be susceptible to several pathogens. Many soil factors including soil type, texture, pH, moisture, temperature, nutrient levels, and ecology affect the activity of soil-borne pathogens.

Disease Management Strategies

Diagnosis and Assessment of Disease Incidence and Severity

Some diseases caused by soil-borne pathogens may be difficult to diagnose due to the absence of characteristic symptoms. Typical aboveground symptoms are wilting, chlorosis (yellowing), and stunting. Root symptoms include swellings (galls, knots, and clubs), lesions, rotting of root tips and fine lateral roots, and loss of root cortical tissue (rat tails). It may be necessary to enlist the help of an experienced pathologist or plant diagnostic clinic for assistance in diagnosis. Crops should be monitored regularly; walk in a zigzag pattern across the field to maximise the chance of finding problems. Written records of disease outbreaks should be kept for each field because soil-borne pathogens persist in soil for long periods of time. Crop health assessments should include disease incidence, disease severity, and crop loss estimates—all key factors when considering control strategies.

Seed Selection and Storage

Planting pathogen-free and high-quality vegetable seed is a critical first step in managing diseases. This is particularly important for vegetatively propagated material such as tubers, bulbs, and slips. True seed can also contain pathogens, however, and all can serve as the source of entry of pathogens into new areas. Good management practices and seed storage prior to planting are essential for reducing diseases caused by soil-borne pathogens such as Rhizoctonia (stem canker/black scurf) and Fusarium (dry rot), and secondary bacterial soft rot in cut potato seed. In conventional production, seed is often treated with chemical fungicides to reduce seed and seedling losses.

Variety Selection and Grafting

Many cultivars (varieties) of vegetables have been developed that are resistant or tolerant to specific diseases. The terms resistance or tolerance refer to a plant's ability to overcome, to some degree, the effect of the pathogen. Typically, varieties labelled resistant to a particular disease have a higher level of resistance than those labelled tolerant. No variety is resistant or tolerant to all pathogens. For example, tomato hybrids acceptable for the fresh market tomato industry with resistance to fusarium wilt, fusarium crown rot, root-knot nematodes, and verticillium wilt (but not other diseases) are widely available commercially. Therefore, resistant varieties should always be used in combination with other management practices for a complete management programme.

In recent years, grafting of solanaceous (tomato, eggplant, pepper) and cucurbit (melon, cucumber, watermelon) scions onto compatible disease-resistant root-stocks has become a means to raise popular varieties susceptible to certain diseases in fields infested with those pathogens. Grafting can be used to manage soil-borne diseases such as bacterial wilt and root knot nematode of solanaceous vegetables and fusarium wilt of cucurbits. While grafting can protect vegetable crops from some soil-borne diseases, root-stocks are not resistant to all diseases. Knowledge of the causal agents of crop problems is necessary to select appropriate resistant varieties.

Organic Matter Amendments

The addition of organic matter such as cover crop green manure (single and mixed species), seed meals, dried plant material, good quality compost, organic waste, and peats can aid in reducing diseases caused by soil-borne pathogens. Several studies have shown that organic matter amendments can be very effective in controlling diseases caused by pathogens.

Organic matter improves soil structure and its ability to hold water and nutrients; it also supports micro-organisms that contribute to biological control. Our study has shown that mixed-hay cropping during the transition periods can enhance soil suppressiveness to damping-off caused by Pythium and Phytophthora; in addition, although compost amendments applied during transition can improve crop vigour by significantly enhancing soil fertility, their effects on soil-borne diseases are not predictable when transitioning to certified organic production.

Organic matter amendments have great potential. However, they sometimes can cause inconsistent control, increased disease severity, and phytotoxicity. Correct management of crop residues and wastes is necessary to avoid phytotoxic effects. This can be achieved by optimising application rates and proper timing between organic matter application and planting of the vegetable crop.

In the early stages of decomposition—especially when the available oxygen is low (as in saturated soil)—crop planting should be avoided, or at least delayed, to avoid phytotoxicity and/or diseases caused by Pythium and related pathogens. Phacelia, Austrian winter pea and lana woollypod vetch cover crops can increase the number of sclerotia of Sclerotinia minor. Vetch is a known host of root-knot nematode (Meloidogyne species) and also might increase populations of Pythium and Rhizoctonia which cause damping-off. Although oilseed radish could be a potential trap crop for cyst nematode (Heteroderaspecies), as a cover crop it is a host of root-knot nematode and the clubroot pathogen (Plasmodiophora brassicae). Grower management of brassica cover crop residues could greatly affect biofumigation effectiveness. For maximum effect, residues need to be completely shredded and immediately incorporated into moist soil.

Propagation, Nursery, and Crop Establishment

Vegetables can be either transplanted or direct-seeded. Transplants can escape damping-off and certain root rot problems if sterilised soilless growing medium is used and seedlings are properly managed. Overwatering and temperature extremes should always be avoided during transplant production. For direct-seeded vegetables, good soil drainage through proper soil preparation helps reduce soil-borne disease. In heavier soils, shallow planting is recommended to reduce the risk of damping-off disease problems. Raised beds are ideal for vegetable production (of non-vining cucurbit crop types). They help prevent damage from soil compaction and flooding. Optimum plant spacing helps to increase airflow around the plants and minimise relative humidity and leaf wetness, which many pathogens need to infect foliar plant tissue.

Organic mulches aid in disease prevention by reducing direct contact between soil and the above-ground portions of the plant. Many soil-borne pathogens infect the plant from rain-splashed soil on the lower leaves. Irrigation management is clearly an important factor for soil-borne disease control. Regardless of the irrigation method, timing, frequency, and amount are critical. Over-watering in poorly drained soils must be avoided to prevent soil-borne diseases and seed decay. In research on irrigation methods, lettuce drop incidence was significantly lower and yields were higher in plots under sub-surface drip irrigation compared with furrow irrigation. The selection and application of fertilizers can significantly influence disease development. For example, liming the soil to raise pH levels can reduce symptom expression for clubroot disease of crucifers and help manage bacterial wilt of tomato. The combination of high-nitrate, low-ammonium fertilizers and application of lime can also provide significant control of fusarium wilt and fusarium crown rot of tomato.

Sanitation

Many pathogens survive between crops in or on the residue from diseased plants, so it is important to remove or destroy as much of the old plant debris as possible. Tillage procedures should also reduce plant residues left from previous crops. For example, when lettuce drop or white mold is observed in a production field, the diseased plants and soil immediately surrounding them should be removed, to reduce the number of sclerotia (long-term survival structures of the pathogen) in the soil. Weeds should be eliminated as they may harbour pathogens or serve as hosts; they also increase the relative humidity around the crop plants, favouring disease development. Equipment should be cleaned by power washing between fields. For crops such as tomatoes that are handled often for procedures such as training, pruning and harvesting, hands and tools should be cleaned regularly. Sanitation measures are especially important in greenhouse crop production systems. Greenhouse vegetable crops are to a large extent produced in soilless cultures. Soilless cultures were initially used as a method to eliminate soil-borne pathogens. However, vegetable crops in soilless cultures are also affected by the same soil-borne pathogens (Pythium, Phytophthora, Fusarium, and Rhizoctonia) as crops in soil. The advantage of soilless cultures is that the roots are confined to a small volume of medium or water, and that pathogens can be controlled relatively easily. The growing medium can be replaced or sterilised after usage. The source water, recycled nutrient solution, and run-off from a bag culture in a re-circulating system can be treated. There are many different methods available : heat pasteurisation, UV radiation, chlorine treatment, slow sand filtration, ultrafiltration, micropore filtration, and biofiltration. Work surfaces, tools, gloves, boots, hoses, walkways, carts, tractor tires, plows, tanks, and water supply lines in greenhouses, shade houses, storage rooms, and field operations need to be cleaned and disinfested with a disinfectant (such as chlorine materials and hydrogen peroxide) or steam. Surface disinfectants are allowed with restrictions.

Soil Solarisation

Soil solarisation is a preplant and hydrothermal procedure that uses transparent film to capture solar radiation in the soil. The plastic sheets allow the sun's radiant energy to be trapped in the soil, heating the upper layers. Solarisation has gained popularity among greenhouse and high tunnel (hoop-house) vegetable growers.

Application of soil solarisation for open field soil disinfestation has been limited to small areas in geographic regions with a period of high temperature and abundant sunshine. Solarisation during the hot summer months can increase soil temperature to levels that kill many important soil-borne fungal and bacterial plant pathogens, including Verticillium dahlia, certain Fusarium spp., Sclerotiniaspp., Agrobacterium tumefaciens, Streptomyces scabies, nematodes, and weeds.

Anaerobic Soil Disinfestation

Anaerobic soil disinfestation works by creating anaerobic soil conditions by incorporating easily decomposable organic materials (wheat bran, molasses, rice straw, and rice bran, for example) into soil that is irrigated and subsequently covered with plastic film. The advantage of anaerobic soil disinfestation over soil solarisation is that the method does not require high solar radiation so it can be applied to cloudy areas or periods of low sunlight, thus a growing season is not lost.

In practice, 1–2 Mg of organic material are added to each 1,000 m2 (roughly 5–10 tons acre-1) of soil treated. Anaerobic soil disinfestation using wheat bran or molasses proved to be effective against a wide range of soil-borne pathogens, including Fusarium oxysporum f. sp. lycopersici, Verticillium dahliae, andRalstonia solanacearum, as well as the nematodes Meloidogyne incognita and Pratylenchus spp.

Crop Rotation and Inter-cropping

Crop rotation is the foundation of disease management for many pathosystems and a very important tool in organic agriculture. Alternation with hosts that do not support growth and reproduction of the pathogen reduces the amount of soil-borne inoculum. A good rotation arrangement is to treat members of the same plant family as a group and rotate based on groups rather than individual crops. For example, lettuce should be rotated with cucurbit, tomato, or cole crops but not other types of lettuce, chicory, or endive. It is important to note that one difficulty in using crop rotation to control some diseases is the limited availability of crops for rotation. For example, Sclerotinia on cabbage is difficult to manage with rotation because dicot plants are hosts of Sclerotinia and the pathogen can survive for long periods between host crops. Peppers should not be rotated with cucurbits, as both are susceptible to phytophthora blight. Green manure or cover crops can be part of the crop rotation. When selecting cover crops on the basis of disease suppression, caution should be taken because the effects can be cultivar specific. Broccoli (Brassica oleracea), used as a green manure, has been shown to reduce the number of S. minor sclerotia in soil (Hao and Subbarao, 2003).

Intercropping is defined as the cultivation of a mixture of two or more crops together in the same area. There are various types of intercropping (row intercropping, strip intercropping, mixed intercropping, and relay intercropping), each of which may affect soil-borne pathogens differently. Intercropping systems can be characterised according to the degree to which roots of different crop species interact, which is determined not only by the intercropping system but also by the root architecture of each of the crops in the mixture.

Disease-suppressive mechanisms such as host dilution, allelopathy, and enhanced antagonistic populations play crucial roles for soil-borne disease reduction in intercropping systems. Reduction in plant pests and weeds has also been reported. Intercropping of compatible plants also encourages biodiversity by providing a habitat for beneficial insects and soil micro-organisms.

Materials

Applications of copper products may provide some protection against the foliar phase of phytophthora blight, a soil-borne disease. However, over-application can lead to copper accumulation in the soil, contamination of run-off water, and toxicity to non-target organisms. Biorational products represent an important option for the management of plant diseases. Streptomyces griseoviridis and emdash; commercially known as Mycostopandemdash; and species of Trichoderma are among the micro-organisms most commonly used for biocontrol in soil.

Both agents are used for the control of root diseases caused by Pythium, Phytophthora, Rhizoctonia and Fusarium. The strain T-22 of Trichoderma harzianum is effectively used for seed treatment. Kodiak (Bacillus subtilis GB03) and Actinovate (Streptomyces lydicus) are also used for seed treatment. Contans® (Coniothyrium minitans) has been effective in reducing the incidence of lettuce drop by as much as 50 per cent. It is also effective for white mold. Biochemical biopesticides (extract of Chenopodium and thyme oil, for example) can be used for soil-borne disease management. Microbial inoculants to promote plant growth and enhance resistance have been developed and field tested for soil-borne disease management.

HEAT TREATMENTS TO MANAGE SOIL-BORNE PESTS

The objective of heat treatment of soil is to reduce the numbers of weeds, insects, and disease-causing organisms in the soil and thereby promote plant vigour, increase yield and quality and decrease the need to use pesticides later in the production cycle. Soil treatments complement good disease control but do not replace the use of disease-free seed, cuttings, and transplants nor lessen the need for proper sanitation practices. Equipment is available to generate steam or aerated steam in the field or steam generating facilities used to heat greenhouses or other structures can be modified to treat soil brought near the facility.

Timing

The best time to treat is in the fall while soil temperatures are above 55F (13C) at a 6 inch (15.2 cm) depth and soil moisture is 50-85 per cent of field capacity. Wet soil reduces the effectiveness of treatment because heat does not penetrate well. On the other hand, pests in dry, cold soil are usually in a resting state very resistance to treatment.

General Procedure in Plant Beds

In the fall, disk or rototill the soil and allow some time for the organic debris to rot, usually at least 7-10 days. Soil moisture should be 50-85 per cent of field capacity and soil temperature above 55°F (13°C) at a 6 inch depth (15.2 cm). Either inject the steam directly into the soil or cover the area with a canvas tarp and pipe the steam under the tarp.

General Procedures for Potting Soil

Soil should be on a surface that will allow the soil to be moved without re-contaminating it with untreated soil (on a wooden, metal, asphalt or concrete surface or plastic tarp). The layer of soil must be no deeper than 6 inches (15.2 cm) at any point. Cover the pile with a canvas tarp and pipe the steam under the tarp.

Monitoring the Temperature

Place a thermometer near the edge of the area or pile to be treated and another in an area away from the edge but close enough to be able to be read as the heating continues. Once the desired temperature is reached on the edge (where heat will be lost quickest) and within the area treated, THEN begin timing the treatment.

Re-contamination

Pests inadvertently added to treated soil have no competition and may cause more damage than they normally could in untreated soil. Disinfest all tools, flats, and equipment that will be used to handle the treated soil. Do not walk or drive on treated soil.

Chapter 17

Farming Systems for Managing Soil-borne Pathogens

BACKGROUND

Edible ginger (Singiber officinale) production in Fiji dates back to the 1950s when a small area was grown in the Suva peninsula with produce exported to New Zealand. In the 1960s the production dramatically increased as Fiji started exporting fresh ginger to North America and new production areas were identified. Production increased to 5,000 metric tonnes in the 1980s and stayed around that figure for some time until 1999 when the North American market collapsed due to competition from China. The industry has since recovered and restructured with an emphasis on exports of fresh immature ginger and the entry of ginger processors that preserve immature ginger in brine or syrup for export markets. In exports alone, the industry was valued at FJD$4.2 million in 2004 when this project first commenced activities. Ginger is planted in Fiji between September and early November, and immature (green) ginger is harvested January-March (70 per cent of the crop) and mature (brown) ginger in June-July. Ginger is also harvested for 'seed' which provides the vegetative planting material for next season's crop. Ginger is mainly grown in the wetter areas of Viti Levu (Suva/ Nausori/Navua) and predominantly by Fijian small-holders as an important cash crop. Exporters and processors contract growers to produce an agreed quantity of good quality ginger at a set price.

The past 15 years has seen a decline in ginger production and greater wastage due to soil-borne diseases mainly occurring during the wetter, warmer months. Several recent surveys, Hort Research/Fiji Government in Fiji and Biological Crop Protection/Horticulture Australia in Australia, provide convincing evidence that soil-borne diseases are the major problem. A major constraint to production in Fiji is the soil-borne pathogen, Pythium myriotylum (Pm), which infects seed-pieces and causes rot and early death of plants.

Pm was killing up to 30 per cent of plants during surveys undertaken during 1997/98 but is capable of completely destroying the entire crop. In Australia the major soil-borne pathogen of edible ginger was widely regarded as Fusarium oxysporum f. sp. singiberi (Foz), until the wet summer of 2007/08 when Pythium myriotylum was isolated on a few farms. Both are now responsible for extensive rhizome rots. The epidemiology of both Pm and Foz is also influenced by nematode damage to roots and rhizomes, which can exacerbate disease development. Annual farm-gate losses due to soil-borne pathogens were estimated in 2004 at FJD$530,000 for Fiji and AUD$990,000 for Australia, and the situation was deteriorating.

These pathogens limit production and affect rhizome quality, which in turn can decrease profits and destabilise the market. Furthermore pressure is placed on 'seed' beds as a source of material to satisfy contracts. If the best material is sold and pathogen infected material used for planting, then a downward spiral begins. If serious enough, pathogens can even lead to market failure. It was important that the disease/pest cycle be broken.

A review of the ginger industry was undertaken by the Fijian Ministry of Agriculture in conjunction with a major growers' meeting during 2004.

The major issues and priorities, as listed at the meeting, included :

- Increasing threat of disease – production losses/poor quality/lack of seed ginger
- Increasing cost of production under current farming system
- Scarcity of labour
- Low labour productivity
- Lack of credit facilities
- Environment degradation
- Urban drift – farming no longer an attractive option for the younger generation
- Institutional problems.

Despite these concerns the industry concluded, "Fiji ginger is known for its taste but can no longer be complacent in terms of quality and price. There is great potential for expanding the processing sector and fresh immature ginger may become a specialty product in the hands of a few exporters. Fiji is capable of producing quality ginger and at a competitive price providing the problems highlighted above are addressed effectively and in time." It is interesting to note that the priorities identified by the Fijian ginger industry are almost identical to the Australian ginger industry, and in the same order. ACIAR project PC/2004/049 attempted to address the priority issue of soil-borne diseases and the use of more sustainable production practices should also address the issue of environmental degradation. Creation of a more profitable ginger industry should have flow on benefits for regional communities as well as businesses in the larger towns.

Research Strategy

The research strategy for the project involved developing a farming systems approach to ginger production using pathogen-tested planting material, bedding/mounding, reduced tillage, crop rotation and the use of mulches/organic amendments to improve soil physical and chemical properties and suppressiveness to soil-borne pathogens. The key was to achieve an active soil biology that was antagonistic to fungal pathogens, enhance populations of free-living nematodes to levels that maintain predators of nematodes (particularly fungal predators), and limit nitrogen to levels that encourage predatory activity. Improvements in soil health status were confirmed with some simple measurements of microbial activity, free-living nematodes, as well as measures of nutrient status and soil organic matter. Combining this with yield data and disease incidence enabled us to determine whether we improved the soil's suppressiveness to key soil-borne pathogens and restored its biology. This knowledge immediately translates into practical recommendations for the most appropriate farming systems to use for ginger production in Fiji and Australia.

This project drew on findings related to ginger marketing and economics of the ACIAR project ASEM 2003/069 Policy options for improving the value of land use in small-holder Fijian Agriculture. This project undertook a small study to look at the economics and marketing of ginger during the second year of its implementation through a Masters Research project by Waisiki Gonemaituba at the University of Queensland. This project also had links with other ACIAR projects in the Pacific dealing with IPM of vegetables *e.g.* CP 2004/063 Integrated pest management in a sustainable production system for Brassica crops in Fiji and Samoa.

A farmer participatory research (FPR) approach was used at 3 farmer trial sites in Fiji (Waibau, Navua and Veikoba) and 2 trial sites in Australia (Kandanga and North Arm). This research was also linked with the Fijian Ministry's extension service and a senior officer was devoted full-time to ginger and six other officers were located in the major ginger producing areas with 30 per cent of their time responsible for providing support to ginger growers in the area. In addition the exporters/processors, who have a vested interest in the supply of quality material, have liaison officers who work with farmers as well as the Ministry's Extension Division. These people knew the growers and had the necessary social-cultural skills to help drive the FPR model. FPR, both in Fiji and Australia, had the advantage that industry champions were identified so the project was tailored to more effectively deliver agreed outputs.

The Fijian government targeted support for export crops as very high priority and there was strong support and commitment for the ginger project at all levels. In this context the Ministry had a policy in place to set up model farms to not only provide demonstrations of best farming practice, but to provide industry support in the form of clean planting material.

Aim and Objectives of the Project

The overall aim of the project was to improve profitability (yield and quality) of ginger in Fiji and Australia through better management of soil-borne diseases. This was to be achieved through the following three objectives :

Develop practical systems for production of clean planting material :

- Overcome the practical problems involved in maintaining blocks of clean planting material on ginger farms in Fiji and Australia, and demonstrate the value of clean planting material for managing soil-borne diseases.

Develop soil management systems to improve soil health and reduce losses from soil-borne diseases :

- Improve the suppressiveness of ginger-growing soils to soil-borne pathogens (particularly Pythium myriotylum, Fusarium oxysporum f.sp. singiber, Meloidogyne spp. and Radopholus similis) by incorporating rotation crops (*e.g.*, taro, cassava, brassicas, legumes, grasses), minimum tillage, mulches and organic amendments (*e.g.*, rice husks, sugarcane trash, mill mud) and plant defence activators (*e.g.*, silicon) into the ginger farming system.

Implement improved ginger production systems :

- To communicate results at grower and community forums and co-ordinate activities with local industry groups.
- To develop a plan for engagement and adoption of the recommendations made by the project team through a consultative process. The network of contracted growers is the focus of activities.

The outcome of these studies will be to increase the production of high-quality ginger in Fiji and Australia. They will lead to the development of farming systems for ginger that improve soil health, sustain productivity and increase profitability.

Develop Practical Systems for Production of Clean Planting Material

The methods for tissue culture propagation of ginger are well established and in routine use at Maroochy Research Station and Koronivia Research Station. Small trial plots of tissue cultured material were established in Fiji and Australia and seed-pieces harvested from these plots were used to establish nursery beds for the continued production and supply of superior quality 'seed' material.

Nursery sites were identified that have not grown ginger or used water from dams for past ginger production. The nursery sites were also managed as if they were quarantined from the rest of the ginger industry. Sustainability of such schemes will depend on clear demonstration of benefits and results from our studies have highlighted the value of clean 'seed' as compared to the 'seed' currently in use. This is particularly important in Fiji to stop the practice of buying second-grade seed or in using market reject material as planting material.

Seed treatments consisted of allowing the seed to suberise for 4-5 days under ambient temperatures vs seed that had been cut and immediately planted, as well as hot-water treatments where seed was dipped in water maintained at either 48°C for 20 minutes or 54°C for 15 minutes.

DEVELOP SOIL MANAGEMENT SYSTEMS

Observations on Crop Losses and Disease Etiology

Fijian fields were checked for signs of Pythium Soft Rot and plants with yellow or dead shoots were inspected. If symptoms were typical of soft rot, rhizomes were collected and the diagnosis confirmed by isolating the pathogen in the laboratory. Ginger brought to the processing factory in February 2007 was also checked for soft rot and other pests and diseases. Twenty consignments of ginger from various parts of Fiji were carefully inspected, the amount rejected for various reasons was recorded and pathogens associated with disease symptoms were confirmed in the laboratory. Since most Fijian crops are harvested in February as immature ginger, observations on the development of rhizome rot later in the season were limited to crops retained for planting material.

The spatial distribution of the disease was monitored in eight fields (four near Veikoba and four in the Waibau region) that ranged in area from 48 to 240 m2. All crops were disease-free in February 2008 and were therefore retained to provide seed later in the season. Fields were inspected in March, April and May 2008 and the location of all plants showing disease symptoms was marked on a map. Final observations were made when crops were harvested for seed during late July and early August 2008 and seed losses were estimated in consultation with the grower.

Pythium Soft Rot

Pythium Soft Rot had never been observed on ginger in Australia until late December 2007, when a severe outbreak occurred on two farms in Queensland that were 28 km apart but owned by the same family. Initial observations were made soon after the disease was reported by the grower and disease spread was then monitored in four fields that had distinct patches of diseased ginger surrounded by green, healthy plants. Three of the fields were at Eumundi and one at Kandanga, and they varied in size from 1.4 to 3.6 ha. Datum plots (10 m long with three rows of ginger contiguous along beds 1.9 m wide) were marked out in each field and disease severity was rated every few days as follows : 0 = no yellow and necrotic shoots in the datum plot; 1 = <2 per cent; 2 = 3–5 per cent; 3 = 6–10 per cent; 4 = 11–33 per cent; 5 = 34–67 per cent; 6 = 68–100 per cent diseased shoots in the datum plot. The rate of disease spread was measured by first marking the edge separating diseased and yellowing shoots from green and healthy shoots. The daily rate of spread was calculated by measuring the distance the diseased edge had moved between recording dates. Fields were monitored every few days from mid-January to early February 2008, and final observations were made just before harvest of immature ginger in late February 2008.

Rainfall and Temperature Data

Since outbreaks of soft rot are known to be associated with hot and consistently wet conditions, temperature and rainfall data were collated for two sites : Koronivia Research Station, Nausori, Fiji and Maroochy Research Station, Nambour, Queensland, Australia. Both stations are located within 20 km of the major ginger-growing areas in each country and have a climate that is broadly similar to those areas. The records date back to 1973 for Nausori and 1952 for Nambour. Long-term averages for December–May and data from the 2006–07 and 2007–08 ginger-growing seasons were retrieved for rainfall, the number of rain days and minimum and maximum temperature.

Isolation and Identification of Pythium

Ginger rhizomes showing disease symptoms were washed thoroughly in water, dipped in ethanol and flamed briefly before tissue from the leading edge of lesions was transferred to media that were selective for Pythium. In Fiji, isolations were done on cornmeal agar with carbendazim (2 mg/mL), ampicillin (250 mg/mL), rifampicin (10 mg/mL), pimaricin (5 mg/mL), and pentachloronitrobenzene (100 mg/mL), while a medium containing penicillin, polymixin and pimaricin was used in Australia (3P medium of Eckert and Tsao 1962). To check for fungal pathogens other than Pythium, tissue was also plated onto potato dextrose agar containing streptomycin (120 mg/mL).

Pythium isolates were identified using methods outlined by Van der Plaats-Niterink (1981). Growth rates at various temperatures were determined on potato carrot agar while sporangia, antheridia, oogonia and oospores were produced by transferring the fungus to autoclaved grass leaves floating in sterile water. The identity of isolates used in pathogenicity tests was confirmed by sequence analysis of rDNA internal transcribed spacer (ITS) regions containing ITS1 and ITS2 and intervening 5.8rDNA using the primer pair ITS1/ITS4.

Pythium Soft Rot Pathogenicity Experiments

Experiments were set up in Fiji and Australia using isolates of P. myriotylum from each country. Inoculum was prepared by placing 50 g sorghum seed in 250-mL Ehrlenmeyer flasks, soaking the seed overnight in water, pouring off the excess water, autoclaving twice on successive days and then inoculating the seed from cultures on agar. The inoculum was used after the oomycete had fully colonised the substrate (usually 7–10 days at 25°C). Plants were grown in pots containing steam-sterilised potting mix (3 sand : 1 peat) and were inoculated 2–5 months after a seed piece had been planted in each pot. After 4 g of inoculum was added, the soil was kept moist for 4 days and then a wet/moist cycle (4 days wet and 3 days moist) was imposed by placing a saucer under the pot for 4 days and removing it for 3 days. At the end of each experiment, small pieces of tissue from rotting rhizomes were transferred onto a selective medium to check for the presence of Pythium. In both experiments, the level of disease in each pot was assessed periodically by rating aboveground symptoms using the following scale : 0 = healthy plant; 1 = some leaves starting to turn yellow; 2 = plant alive, but shoots either totally yellow or dead; and 3 = all shoots dead.

Plant Parasitic Nematodes

Surveys of plant parasitic nematodes that could also be responsible for yield losses were conducted in farmers' fields from all the major ginger growing areas in Fiji. Soil was collected from around the roots and rhizomes, mixed gently and a 200 ml sample was placed on a standard extraction pan for 2 days. Nematodes recovered by sieving twice on a 53 μm sieve were counted.

A total of 20 consignments of ginger were checked at the Frespak factory at Lami by inspecting ginger on the conveyor belt after washing and in the reject bins. About 100 rhizome pieces in each consignment were carefully inspected and lesions possibly caused by nematodes were teased out under water to check for the presence of root-knot nematode (Meloidogyne spp.) and burrowing nematode (Radopholus similis). The washed weight of each consignment and the weight of ginger rejected were obtained from factory records.

Radopholus. Relatively small (5-15 mm diameter), light brown or water-soaked lesions generally extending only 1-2 mm below the surface. Meloidogyne. Slight eruptions of tissue usually near the base of developing buds.

Burrowing Nematode Pathogenicity Experiment

On 12 September, 2008, a potting medium consisting of 2 parts washed river sand and 1 part commercial potting mix was heated to 70°C for 30 minutes. Twenty 4L pots were then filled with this medium and a metalaxyl-treated seed piece of ginger (Z. officinale) was planted in each pot. Pots were transferred to a glasshouse and after 6 weeks, when plants had shoots up to 10 cm long, half the pots were inoculated with 1,500 Radopholus similis. The nematode was originally obtained from a ginger farm at Veikoba, Fiji and had been multiplied in the laboratory on sterile carrot tissue.

On 9 February, and 20 March, 2009 (15 and 20 weeks after pots were inoculated), the total number of shoots and the number of yellowing or dead shoots in each pot was recorded. At each harvest date, the above-ground biomass in five inoculated and five control pots was measured and seed pieces and newly-developing rhizomes were then removed from the pots for assessment. Roots were chopped into segments <1 cm long and mixed with the soil, and a 200 ml sample (about 220 g) of soil and roots was spread on a Baermann tray. After 2 days, nematodes were recovered on a 38 μm sieve.

Portions of seed pieces and rhizomes that showed symptoms possibly caused by R. similis were assessed by removing small pieces of tissue (about 1 mm3) from affected areas, macerating them in water and checking for nematodes after 24 hours. Discoloured tissue was also checked for fungal pathogens by placing small pieces of the tissue onto potato dextrose agar (PDA) or corn meal agar with carbendazim, ampicillin, rifampicin, pentachloronitrobenzene and pimaricin (CARPP) and observing plates after 24 and 48 hours. The nematodes present in larger pieces of discoloured or rotting tissue (*i.e.* 0.5 - 4 g) were recovered by macerating the tissue in 200 ml of water in a blender and then placing it on a Baermann tray for 24 hours. This maceration technique was also used to recover nematodes from the remaining part of each seed piece or rhizome. In cases where nematodes were extracted from several pieces of tissue from the same seed piece or rhizome, the counts were added to obtain the total number of nematodes.

Developing Farming Systems to Control Soil-borne Pathogens

In Fiji a field trial at Waibau was established in September 2006 to examine the effect of nitrogen (N) on Pythium Soft Rot. The experiment consisted of four rates of poultry manure (0, 5, 12.5 and 17.5 t/ha) applied before planting to four replicate plots (each 20 m2) set out in a randomised block design. In Australia the trials were larger, with more treatments and larger blocks to accommodate mechanised operations. An experiment was established to examine the influence of cover cropping, pasture leys, organic amendments and minimum tillage in a ginger farming system over 2 crop cycles.

Soil textural and chemical analyses were made from soil sampled to a depth of 10 cm and major and minor elements were measured using a commercial nutrient analysis service in Australia (Incitec Pivot, Werribee) or the Soil Testing Laboratory at Koronivia Research Station.

The presence of soil-borne pathogens was confirmed by collecting soil samples from a depth of 0-10 cm on 6 April, 2006. The soil was mixed gently, spread on a standard nematode extraction tray and after 2 days, nematodes were recovered by sieving twice over a 38 μm sieve. Bioassays in which tissue-cultured ginger plants were planted into 1 L of field soil diluted either 1/10 or 1/100 with autoclaved soil and grown in a glasshouse for up to 8 months to detect presence of fungal pathogens.

The basic experiment consisted of 5 cropping/tillage treatments × ± poultry manure amendment × 4 replicates. The cropping tillage treatments were set out in a randomised block design, with the amendment as a split plot.

The main treatments were :

- Conventional tillage with continuous cropping and a conventional tillage ginger crop,
- Minimum tillage with continuous cropping and a conventional tillage ginger crop,
- Conventional tillage with a summer crop only and a conventional tillage ginger crop,

- Minimum tillage with a summer crop only and a minimum tillage ginger crop, and a four-year pasture ley further split into (ECT) conventional tillage ginger crop and (EMT) minimum tillage ginger crop. A fumigated control (F) was included where conventional tillage summer crops were followed with a conventional tillage ginger crop, however the beds were treated with 500 kg/ha 1,3-dichloropropene (Telone®) 3-4 weeks prior to planting ginger. Field experiments were grown as per commercial practice.

Growth and Yield Measurements of Ginger Crop

Shoot emergence and number of shoots were counted from a 3 m length of bed on the 26 October, 2009, 11 November, 2009 and 3 December, 2009. Rhizomes were harvested from a 1 m section of bed and an estimate of yield extrapolated from the weight of uninfected rhizomes (t/ha). Rhizome that had been rejected due to rot was weighed and expressed as a percentage of the total rhizome harvested from the plot.

The first-late harvest occurred on 4 May, 2010 and again rhizomes were harvested from a 1 m section of bed and both rejected (due to rot) and sound rhizomes were weighed and yields calculated. Root knot nematode damage was slight (none would have been rejected) and was expressed as the percentage of 10 randomly selected rhizomes that showed any signs of damage.

Soil-borne Pathogen Assessments

For the first ginger crop at early harvest on the 12 March, 2008, the ginger in a datum plot 1.15 m long was harvested and the number of rhizomes showing rot symptoms caused by soil-borne fungi was recorded. Laboratory isolations indicated that these symptoms were primarily caused by Pythium myriotylum. However by mid-April yellowing of plants was associated with Foz infection of the rhizomes and it was estimated that another 10 per cent of the trial was affected. On 14 April, 2008 (2 days prior to first-late harvest), three independent observers estimated the percentage dead and severely yellowing shoots in an 18 m datum plot in the centre of each plot. Estimates were then converted to a disease rating of 1 to 6, where a rating of 1 indicating that <5 per cent of shoots showed symptoms and a rating of 6 indicating 45-55 per cent of shoots were affected.

For the second ginger crop and prior to early harvest (31 March, 2010), the percentage of yellowing and necrotic shoots in the entire length of the bed was estimated for each treatment, and plots were rated for disease severity (0 = all shoots green and healthy to 5 = total collapse and death of shoots).

Root knot nematode populations were monitored during the growing season by collecting soil samples prior to planting, at early harvest and 2 days prior to first-late harvest by the method described previously. At each sampling time, 10 cores were collected from each plot to a depth of 0-20 cm using a 2.5 cm diameter sampling tube.

Since root knot nematode was not detected in soil samples collected prior to planting of the first ginger crop, the nematode was inoculated at one point in each plot in an attempt to determine whether some treatments were more suppressive to nematode multiplication than others. Inoculum of M. Javanica was obtained by retrieving eggs from glasshouse-grown tomato plants using NaOCl and 50,000 nematode eggs were then inoculated at a depth of 15 cm, at a point 4.5 m from the end of each plot. Eggs were added on 15 October, 2007, three weeks after ginger was planted but before any emergence had occurred.

On 3 March, 2008 (early harvest), all plants within 20 cm of the inoculation point were dug up and all rhizomes (generally 3-5 pieces) were rated for root knot nematode damage on the following scale : 0 = no damage; 1 = a low level of damage (1-2 areas erupted due to the nematode per rhizome); 2 = moderate damage (3-5 erupted areas); 3 = high (6-10 erupted areas) and 4 = very high (rhizomes unmarketable due to root knot nematode damage). After rhizomes had been removed, the soil in this area was sampled and nematodes were extracted and counted as described previously.

Pythium Suppression Bioassays

In September 2009 (just before ginger was planted and four years after the experiment commenced), soil was collected from each plot and half was irradiated at 25 kGy. Both irradiated (IR) and non-irradiated soils were placed in 2 L plastic planter bags and planted with a 60 g piece of ginger rhizome. Pythium myriotylum was cultured on autoclaved sorghum seed and on 10 December, 2009, when plants had 1 or 2 shoots up to 50 cm long, each pot was inoculated with two Pythium-colonised seeds. The assay was repeated on 19 January, 2010, except each pot received only one colonised seed. The mean maximum temperature during December and January was over 30°C and overhead sprinklers in the screen house provided a watering regime conducive for Pythium infection. After approximately 3 months for each assay, plants were assessed for disease severity on a 0-3 scale where 0 = no disease; 1 = some leaf yellowing; 2 = most shoots yellow or dead; 3 = rhizomes rotted and plants dead. Assessments were based on 17-21 plants for each treatment in the first assay and 11-16 plants per treatment in the second assay. Pythium was confirmed in a random selection of affected rhizomes at the end of the assays.

On-farm Implementation of Improved Ginger Production Systems

Farms for farming systems test sites were selected in consultation with researchers, extension staff, as well as leading exporters/processors. Three sites were chosen to conduct trial work in Fiji, while there were two in Australia.

Key staff met with selected growers before crop planting to discuss pathogen and crop problems and their experience with different practices, to prioritise problems and possible alternatives for testing, and to identify indicators for evaluating progress. At key periods in the crop-cycle, they met again to compare plant growth and disease levels. They also discuss the progress of their experiments. At each meeting they also conducted field visits and discovery exercises to strengthen their ecological knowledge of the crop, pathogens and soil factors. At the end of the cycle, they identified major lessons to be learned in the next crop-cycle and problems for further experimentation or study.

Provision of Tissue Cultured Plants and Clean Seed

A clean-seed scheme operated in Fiji in the past and there is interest from a number of exporters to re-introduce the scheme. Koronivia Research Station also has a tissue culture laboratory that is currently producing tissue cultured ginger for agronomic evaluation. An inspection of the facility and discussions with staff revealed that capacity exists within the Ministry to multiply and harden-off tissue cultured ginger plants.

In Fiji, two sites have been established by the Ministry of Primary Industries for the production of clean seed for use by industry : one is in the Western Division of Viti Levu near Nadi and the other in the Northern Division of Vanua Levu at Seaqaqa Research Station. These areas are declared free of Radopholus similis and helped alleviate short-falls experienced in supply of planting material during the 2008/09, 2009/10 and 2010/11 seasons.

Tissue cultured plants of the improved ginger variety, 'Buderim Gold', were imported to Fiji under permit and established at Koronivia Research Station (KRS). 'Buderim Gold' is a tetraploid white ginger cultivar developed by the former Queensland Department of Primary Industry and Fisheries (QDPI&F) and has significantly larger rhizomes compared to the mother plant, 'Queensland'. After discussions with Fijian fresh ginger exporters on previous visits, they felt a market may exist for this product in Japan and the USA. After discussions with Buderim Ginger Limited, who helped fund the developmental work on 'Buderim Gold', it was decided to release it in Fiji and determine if international markets may exist for a large, white fresh ginger product.

Most of the 'Buderim Gold' plants were established in pots in a glasshouse under shade, but six plantlets were retained for tissue culture multiplication at KRS, while another six plantlets were to be placed in the *in vitro* germplasm collection at the Secretariat of the Pacific Community (SPC).

Two Australian 'seed' producers were each supplied with consignments of 1500 tissue cultured plants over the course of the project. This was part of an ongoing process to ensure that disease-free and pest-free material is introduced into the nursery production system. In addition, the project team continued to test for soil-borne pathogens on these farms to ensure the spread of diseases and pests on planting material is minimised. Demand for clean seed by the ginger industry continues to exceed supply.

Seed Treatments

In Fiji, seed is typically allowed to air-dry under shade for 8-10 days before planting, whereas the economic imperative in Australia is to plant seed as soon as possible following cutting and fungicide treatment. This means that cut surfaces may not be sufficiently suberised before planting and the physical barrier to prevent pathogen invasion may not be fully formed. Pot experiments were conducted in the glasshouse to test the effect of seed drying treatments on germination and susceptibility to pathogens.

When planted into sterilised soil, seed germination was faster when not allowed to dry for 10 days, even though 17 per cent developed Erwinia soft rots in soils that were occasionally watered and this increased to 35 per cent when soils where maintained at field-capacity. Seed that had been allowed to dry for 10 days and had a suberised barrier at the cut end did not develop any rots. These experiments suggest that seed requires a period for suberisation to occur on the cut ends prior to planting to minimise rots and improve establishment of the ginger crop.

In Australia, seed is hot-water treated at 48 °C for 20 min and then dipped in fungicide to control Meloidogyne. This is often restricted to 'seed' being used to establish nursery blocks as the hot-water treatment can reduce yields by up to 20 per cent. This is deemed acceptable because the rhizomes used for next season's planting material must remain undisturbed in beds for up to 12 months.

In Fiji, A small (1.2 m^3) gas-fired vat was loaned by the Ministry of Agriculture for the purposes of hot-water treating ginger seed. A farmer at the Veikoba trial site used his 'standard' method of treating seed which consisted of half-filling the vat with approximately 600 L of water, waiting until the temperature of the water approached 50 °C (varied from 48 °C to 50 °C) and placing a maximum of eight 20 kg bags of seed in the vat and leaving for 15 min. The temperature of the water was 40 °C at the end of this 'standard' treatment. Our project team was allowed to treat one bag of the grower's own seed using the Ministry's recommended treatment of 54 °C for 10 min (this time the temperature was held constant over the treatment period).

A sample of seed was collected and tested. Whereas live nematodes were recovered from seed using the farmer's 'standard' treatment, no live nematodes were detected using MAF's recommended procedure. A significant finding was that Radopholus was detected in grower's seed from 4 of the 8 plots sampled. This is strong evidence that burrowing nematodes are being introduced into prepared plots on seed that has been insufficiently hot-water treated.

Pest and Disease Surveys

The evidence collected from Fiji and Australia strongly supports the association of Pythium Soft Rot with long periods of wet, overcast weather and in situations where the soils remain saturated.

All Pythium isolates obtained from ginger in Fiji and Australia grow well at high temperatures. They continue to grow slowly at 41°C and have radial growth rates on agar of about 30 mm/ day at 37°C. The Australian isolates from ginger have been identified as P. myriotylum and possibly P. singiberis, whereas the Fijian isolates have been identified as P. myriotylum, P. graminicola and P. vexans.

Pathogenicity tests performed in Fiji and Australia confirmed we are dealing with highly pathogenic isolates that are more virulent when soils are saturated and infection does not required damage to the host roots or rhizomes as caused by the nematodes Radopholus or Meloidogyne. It was established that Radopholus is a primary pest causing damage to the roots and rhizome, but economic losses are negligible for the immature harvest. However by the harvest of mature ginger, particularly for rhizome that is used as 'seed', damage can be severe and the wound sites lead to development of rots that can result in extensive losses in the field.

Control Measures for Soil-borne Pathogens

Pythium Soft Rot was widespread and occurred in all of the major ginger growing areas in Fiji. It has resting oospores that can still be infectious even after long fallows. It is also transferred on infected planting material as our studies confirmed. R. similis, on the other hand, occurred on a few farms, and only in the Veikoba region. The nematode was sometimes not detected in soil at the time ginger was planted, but was invariably introduced in planting material. It then multiplied rapidly and was readily detected later in the season. The highest population densities recorded from ginger fields were 170 nematodes/200 ml soil at immature harvest and 410 nematodes/200 ml soil at mature harvest. R. similis was never found in taro or cassava roots, but was occasionally found at low population densities in soil associated with roots of those crops. However, moderate to high populations were invariably found on volunteer ginger, which was always present as a weed in the following taro and cassava crops.

Soils amended with poultry manure and sawdust (PS) and cropped soils that were subject to minimal disturbance (MT) were most suppressive to Pythium Soft Rot and plant parasitic nematodes. Conversely, treatments where the soil microbial community was disturbed or diminished (*i.e.,* fumigated and bare fallow soils) had significantly lower levels of suppressiveness compared to the Crop/MT/PS treatment. These results demonstrate that the way soil is managed influences its suppressiveness to soil-borne pathogens. It should therefore be possible to reduce the impact of these pathogens by modifying the current ginger farming system. Furthermore, in the Fijian ginger farming system, ginger is generally grown every 3 years, with crops of cassava and taro planted in the years between ginger crops. Our results indicate that this is an excellent rotation for managing both Pythium Soft Rot and plant parasitic nematodes. Taro and cassava are poor hosts for both of these pathogens. Even in the Australian context, a 3-year break from ginger is still likely to be useful in managing soil-borne pathogens, as inoculum densities will decline during the break if non-host plants are used, and thus reduce disease severity when ginger is planted.

A framework was developed for the project activities through regular meetings with research and extension staff, as well as external consultants, and was updated annually. This process provided direction to the project and improved our understanding of the soil-borne pathogens affecting ginger production and provided growers with better options for control in two markedly different environments and cropping systems. In addition a survey of the technical efficiency and output losses (pathogens and soil erosion) of Fijian ginger growers was completed by Mr. Waisiki Gonemaituba as part of his Masters studies at the University of Queensland. His study has important policy implications that the Fijian government can use to create a more profitable and sustainable ginger industry.

In terms of industry extension, project staff presented the results of the first, 2-years phase of research completed in Fiji and Australia at a seminar held at Koronivia Research Station on 13 February, 2009. Research staff from the Ministry was present, as were interested SPC plant protection staff and ginger growers. Towards the completion of the project a major workshop was held on 30 July, 2010 at a local processor/exporters premises in Navua and was very well attended, drawing the support of the Minister of Agriculture and the Permanent Secretary of Agriculture, together with the Heads of Departments and influential local growers. The workshop proved to be a catalyst for gaining the support of private business partners in establishing a clean seed scheme in areas declared free from R. similis. It also provided impetus for the printing of information leaflets in the local language and in launching field days and walks for growers in the major production areas.

In Australia, Field Days organised by the Australian Ginger Growers' Association (AGGA) were held in January for each year of the project and results of project activities were presented through field walks, oral presentations and hand-outs. Project updates have also been provided to growers at quarterly AGGA meetings. Also in Australia, with support from Queensland Primary Industry and Fisheries (QPIF) trade and investment officers, a consultative process was commenced to identify pathways by which all stakeholders within the ginger industry supply chain may increase their profitability. The activities undertaken during the reporting period included a preliminary investigation of the ginger industry to better understand issues within the industry, which was followed up with a detailed investigation of the industry to identify market trends and opportunities for the industry. A report was made available to the Australian Ginger Growers' Association on 11 May, 2009. Forward momentum was maintained by the convening of a Ginger Industry Development Workshop on 27 May, 2009 at Maroochy Research Station to develop an action plan that will guide future investment in the industry and foster its development. An industry R&D levy has since received the support of growers and processors and will be administered by the Rural Research and Development Corporation.

Pythium Soft Rot of Ginger

Observations on Crop Losses and Disease Etiology

Our observations, together with the unpublished results of Fullerton and Harris, show that losses due to Pythium Soft Rot occur regularly in Fiji. In 1997–98, rhizome rot destroyed ~30, 75, 30 and 100 per cent of mature ginger crops at four field sites in Fiji and we observed similar losses to seed ginger in 2007–08. We suggest that given the typically wet Fijian climate, rhizome rot is likely to occur on seed crops in most years and will often limit the amount of material available to plant the next ginger crop. Since most Fijian ginger is now harvested in February for the immature market, total losses from the disease are probably not as great as they were 10 years ago. Nevertheless, immature ginger can suffer heavy losses when conditions are unusually wet for a month or two before harvest. Such a situation occurred in the 2008–09 season, as losses of 25–50 per cent were observed in immature crops in the Navua district following 789 mm of rain on 29 days during January 2009.

The fact that rhizome rot occurred in Australia during the time we were studying the disease in Fiji was fortuitous in some ways, as the summer of 2007–08 was the wettest in the Australian ginger industry for more than 30 years. During the 2 months from mid-December 2007 to mid-February 2008, temperatures were high and rain fell almost everyday; conditions that occur regularly in Fiji. Although severe losses from rhizome rot have never previously been observed in Australia, it is possible that the disease occurred at low levels before 2007–08 and had never been reported. Australian ginger growers normally assume that patches of yellowing plants with rotting rhizomes are affected by Fusarium yellow (caused by Fusarium oxysporum f. sp. singiberi), as this disease occurs on almost every farm in the industry. Dohroo noted that wet conditions, high soil water and relatively low soil temperatures were the most important factors influencing the development of rhizome rot in India and our observations suggest that the first two are also important in Fiji and Australia. Continuity of wetness may be as important for disease development as total rainfall, as the disease is clearly associated with free water and saturated soils. Under such conditions, zoospores disperse in soil water and spread the disease to adjacent clumps, although our observations of greater downhill than uphill movement suggest that mass flow of water is also involved. However, we question the contention of Dohroo that Pythium infection is favoured by low soil temperatures. Although this may be true for some Pythium species, P. myriotylum is most active at mid-summer temperatures. Both Fijian and Australian isolates have high optimum temperatures (radial growth rates of 28–34 mm/ day on agar at 37°C) and high temperatures enhance the pathogenicity of P. myriotylum on other crops in Queensland.

Although our pathogenicity experiments were performed with P. myriotylum, we also found other species of Pythium which were associated with the disease. The fact that severe rhizome rot was initially observed on only two Australian farms raises questions about factors that may have caused the disease to express itself on those farms. Rainfall and soil type did not appear to be involved, as farms near the affected Eumundi farm received similar rainfall; the affected farm at Kandanga was slightly drier than other farms in the industry and the disease was not observed on adjacent farms with similar soil types. The most likely explanation is that the soils on the affected farms had deteriorated due to long-term cultivation. The Eumundi and Kandanga farms were first planted to ginger in 1960 and 1978, respectively, and are therefore amongst the oldest in the Australian ginger industry and probably the most intensively cultivated. Variously classified as ferralic nitisols, red ferrosols or krasnozems, these soils were originally relatively fertile and drained readily. However, after 30–50 years of cultivation, it is possible that drainage is now somewhat impeded due to compaction and declining levels of organic matter. Since Pythium is a relatively poor competitor in microbially active soils, a reduction in the soil's biological buffering capacity may also be a contributing factor.

Impacts

Scientific Impacts – Now and in 5 Years

Much of the research being conducted by the project is of significance and should form the basis for a number of publications that will impact on future studies. Serious scientific investigations into the effect of Pythium and Radopholus on ginger production are either lacking or reported in technical reports and newsletters. Few have received scientific scrutiny in peer-reviewed publications. Likewise the concept of minimum-tillage in ginger production and its relationship to suppression of soil-borne pathogens is novel. The findings to be published from these studies will not only be of significance to ginger but to other root and tuber crops.

Research areas that have been investigated during this project and that will have an influence on future work include :

- Etiology and pathogenicity of Pythium Soft Rot on ginger
- Molecular identification of Pythium isolates
- Etiology and pathogenicity of Burrowing Nematode on ginger
- Impact of crop rotation, tillage practice and organic amendments on developing soils suppressive to soil-borne pathogens
- Improved ginger farming systems.

Capacity Impacts – Now and in 5 Years

Two Fijian scientific staff have had opportunities to strengthen their knowledge and skills in plant pathology and plant nematology, respectively. Ms. Mereia Fong was awarded a John Allwright Fellowship to undertake studies for a Masters of Plant Protection at the University of Queensland which was completed at the end of 2008. The degree was a combination of course work and two research projects. The projects investigated the behaviour of the Australian and Fijian Pythium isolates on ginger in *in vitro* studies (under quarantine) and molecular characterisation of these isolates under the supervision of Dr. Liz Aitken. These results were presented as a poster at the 17th Biennial Plant Pathology Society Conference to be held in Newcastle from 29 September to 1 October, 2009.

In addition, Ms. Fong attended the 16th Biennial Plant Pathology Society Conference, as well as a plant bacteriology workshop, in Adelaide from 24-27 September, 2007. This was her first plant pathology conference and reinforced for her the importance of establishing and maintaining contact with a network of other professionals. She also recognised the importance of structuring her research so that it can be published for the benefit of other professionals and senior decision-makers. These are important lessons that were reinforced during the project.

Ms. Una Turaganivalu attended a training programme in practical nematology under the direction of Dr. Graham Stirling, Biological Crop Protection, during June, 2007. She learned the basics about nematode biology and learned how to identify the important plant parasitic nematodes she is likely to encounter. She also learned the correct procedures for collecting soil samples, extraction and counting of nematodes, preservation of nematodes and maintenance of inoculum. Ms Turaganivalu has made great strides in becoming a competent nematologist and we have been very impressed with her work ethic and ability to carry out a research programme in Fiji with growing confidence. Ms. Una Turaganivalu also presented a poster of her pathogenicity experiment with Radopholus at the 17th Biennial Plant Pathology Society Conference. She also attended the 6th Australasian Soil-borne Diseases Symposium held on the Sunshine Coast from the 9 to 11 August, 2010. She again be presented a poster, this time on the role of rotation crops in managing plant-parasitic nematodes on ginger in Fiji. Graham Stirling was the principal conference organiser. Its location on the Sunshine Coast presented an opportunity for local researchers, extension officers and ginger growers to come along. ACIAR project work was represented in five presentations at the Symposium.

The Fijian staff attendance at conferences highlights the importance of establishing and maintaining contact with a network of other professionals. Presentation of their work also demonstrates the importance of conducting their research in such a way that it can be published for the benefit of other professionals and senior decision-makers. Ms. Fong and Ms. Turaganivalu have made great strides in becoming competent scientists and we have developed confidence in their ability to carry out an independent research programme in Fiji.

The project has also been able to supply needed equipment necessary for nematology research as well as securing a room, with shelving installed, so that nematodes can be extracted from samples in a timely manner. Procedures for setting up and taking down samples are working well. During the project repairs were also made to the autoclave and growth cabinet in the Plant Protection Laboratory at Koronivia Research Station, as well as to the glasshouse and soil steriliser. These have greatly improved the capacity of the staff to conduct research at the station and it was made possible through the project.

During the project the capacity to rear pure cultures of Radopholus similis was dramatically improved by the transfer of technology developed by nematologists at QPIF's laboratory complex at Indooroopilly to the Plant Protection Laboratory at Koronivia Research Station. Burrowing nematodes can now be successfully reared on sterilised carrot cultures so that a supply of nematodes is always available for research and teaching. Nematodes, raised in this manner, were used in a pathogenicity experiment described in the previous section and published in an APPS conference proceeding.

In Australia, a notable project development was a demonstrated ability to create the conditions necessary to study Pythium rhizome rot of ginger at our disease nursery site at Maroochy Research Station. Last year we concurred with the findings of Fullerton and Harris that the Pythium/ginger disease system was too difficult to deal with in trial work. The discontinuous distribution of the fungus in the soil, the development of disease 'hot spots' and the particular epidemiological features of the disease (apparently 'turned on or off' by prevailing weather conditions) does not lend itself to experimentation using conventional trial layouts. However this year at Maroochy Research Station, we have demonstrated that it is possible to culture the pathogen and produce inoculum for successful infection of plants (something not practical or welcomed on farms) and we have developed a watering regime that allows the disease to spread in a manner observed during disease epidemics on commercial farms. We can now study the disease under more controlled conditions and examine treatments for disease control with growing confidence.

Finally it was gratifying to witness the initiative taken by one of Australia's leading growers to modify his ginger planter into a double-disc system that facilitates planting into minimum-tilled beds. The efficiencies gained by this system extend to planting into convention-till beds, as less fuel is needed to pull it through the soil and it also enables the grower a larger window for planting as it is less prone to wet soil sticking to the implement and impeding planting operations. This was an important first step in developing an improved ginger farming system.

Community Impacts – Now and in 5 Years

We believe the project has delivered impacts benefiting the scientific and extension community in both countries, improving capacity to deliver scientifically sound information that meets the needs of land-holders and that improves livelihoods of communities through more sustainable and profitable ginger production.

A major focus of the project was centred on experimental work conducted on specific farms with outreach to the local farming communities. Field days and workshops attended by growers, extension staff and research scientists have ensured that project findings have been communicated widely and that innovative farmers have rapidly adopted improved cultural practices.

Economic Impacts

The project has engendered new interest by DEEDI in the economic development of Australian ginger industry and has resulted in surveys of people involved in the ginger supply chain to ascertain opportunities and challenges facing the industry. A report, Queensland Ginger Industry : Overview of the market trends and opportunities has been prepared by QPIF that gives an accurate and up-to-date account of the state of the industry from the perspective of growers, wholesalers and processors.

In Australia, QPIF estimates the current farm gate value of the ginger industry at approximately $15.6 million. Of the 8,000 tonnes harvested annually, 45 per cent goes to the domestic fresh market while 55 per cent is used for processing. It is recognised that the ginger rhizome undergoes substantial value-adding as the commodity is transformed into confectionery products, brewed soft drinks, and therapeutic agents. The value of these products is estimated at over $80 million. Ginger growers and processors are important employers in regional centres and generate export revenue for the country. The producers employ approximately 200 full time staff and 385 casual staff during peak harvesting periods, while the processors employ over 560 staff. This analysis of the Australian ginger industry complements a similar investigation undertaken by Mr. Waisiki Gonemaituba on the Fijian ginger industry. Issues have been identified in both studies which identify strengths of the industry and areas within the supply chain have been selected to focus activities for resolving constraints to supply or for improving demand. A key constraint identified in both studies was that efforts need to be made to reduce losses due to pests and disease.

The study undertaken by Mr. Gonemaituba for his Master of Philosophy in Economics at the University of Queensland promises to deliver economic, social and environmental impacts for those associated with the ginger industry in Fiji. Mr. Gonemaituba holds a senior position in the Ministry of Agriculture and his findings are important as the ginger industry is one of the key industries identified by the Fijian government in its diversification strategy to accommodate the remnants of the withdrawal of the European Union's sugar preferences. There is considerable pressure on small industries such as ginger to search for ways to operate efficiently and contribute to the economy.

In his study, two types of losses (observed and unobserved) are identified for analysis in a single framework. Observed loss refers to ginger rejected in the field after harvesting and does not reach the exporters' warehouses or processing factory due to sub-standard quality caused by diseases and pests. The unobserved loss, on the other hand, is the unattained increase in ginger output due to technical inefficiency which also includes a measure of soil erosion in a stochastic production function. The study aimed to measure the determinants of observed ginger loss and soil erosion. Farm revenue was estimated at F$1.97 million for the 2006 season, while export of fresh and processed ginger (plus local sales) totalled F$14.03 million. While the observed loss was estimated to be F$469,732, the total unobserved loss to the ginger industry was a large F$4.6 million or 32 per cent revenue loss for Fiji's ginger industry. With regard to losses caused by diseases, especially Pythium, it is recognised that it will fluctuate according to weather conditions, while many of the other pathogens can be controlled by attention to planting material and on-farm management practices. Key recommendations are the need for the establishment of a certified clean seed scheme operated by independent industry groups and a soil repair strategy combined with a proper crop rotation regime as soil fumigation is not seen as a viable option. In summary, the study highlights some key variables concerning technical efficiency and calls for an integrated package of policies related to use of best farming practices, land tenure and agricultural extension and support services for sustainable agricultural growth.

Social Impacts

Social benefits will include the reduction in the amount of pesticide needed (through greater reliance on crop rotations, tillage practices and organic amendments), making agricultural practices safer for producers. There are community expectations that agricultural production will not impact on the safety of farm workers or the surrounding environment, which this project will help to achieve. The empowerment achieved by grower participatory approaches used in the project will enable members to better manage their resources for greater productivity by reducing losses to diseases and pests. From an Australian perspective, minimum tillage is a new concept for the ginger industry. At the most basic level, reducing the number of tillage operations will reduce fuel usage and save money. However the biggest gains will be made by leaving beds as undisturbed as possible, even while continuously cover cropping, thereby improving soil biology and hopefully creating soils that are more suppressive to soil-borne pathogens. From a rural community's perspective, such practices are meant to make farm enterprises in the district more profitable and sustainable with added benefits of reduced erosion, dust and noise.

Environmental Impacts

The project aimed to protect the environment surrounding agricultural production areas by reducing the farming impact on the off-farm environment. To reduce these risks the project aimed to develop more sustainable, yet profitable, management practices. These included optimising the use of organic residues (both green wastes and organic amendments) to develop soils that were better able to suppress soil-borne pests and diseases, as well as retain water and nutrients. A major gain, both in Australia and the Pacific islands, is reduced soil erosion and nutrient run-off related to both the use of appropriate plants and to increasing the stability and resilience of soils.

Communication and Dissemination Activities

As described in Section 6, project staff presented the results of the first, 2-years phase of research completed in Fiji and Australia at a seminar held at Koronivia Research Station on 13 February, 2009. Research staff from the Ministry was present, as were interested SPC plant protection staff and ginger growers. Towards the completion of the project a major workshop was held on 30 July, 2010 at a local processor/exporters premises in Navua and was very well attended, drawing the support of the Minister of Agriculture and the Permanent Secretary of Agriculture, together with the Heads of Department and influential local growers. The workshop proved to be a catalyst for gaining the support of private business partners in establishing a clean seed scheme in areas declared free from R. similis. It also provided impetus for the printing of information leaflets in the local language and in lauching field days and walks for growers in the major production areas.

In Australia, Field Days organised by the Australian Ginger Growers' Association (AGGA) were held in January for each year of the project and results of project activities were presented through field walks, oral presentations and hand-outs. Project updates have also been provided to growers at quarterly AGGA meetings.

MANAGEMENT OF SOIL-BORNE PATHOGENS

Soil is a reservoir for many plant pathogens and plants are under constant attack by these soil-borne organisms. If conditions become favourable for infection, plants will develop disease. Population levels of soil-borne pathogens, including bacteria, fungi, nematodes, and some viruses can be reduced in the soil by appropriate treatments.

Learn which pathogens attack the crop to be grown. Examine the crop regularly, at least weekly, for symptoms of disease or signs of pathogens. To monitor for root diseases, select a few plants from different locations and remove plants from their containers or gently scrape or wash away soil. Examine roots and crowns for browning, softness, or other early indications of disease. Also look for discoloured or wiltedplants and fungal growths aboveground, which may indicate more advanced stages of disease. Test kits are available for detecting Pythium, Rhizoctonia, and Phytophthora fungi infecting greenhouse and container-grown nursery plants. However, keep in mind that some test kits do not specifically test for pathogenic species; in such cases, non-pathogens could result in a false positive result. Use test kits in combination with other information to make good pest management.

Understand the conditions and practices that promote disease and regularly examine for and remedy disease-promoting conditions and practices. Poor sanitation, inadequate drainage, and improper irrigation are the primary conditions that promote diseases of roots. Removecrop residue and old or low-quality plants that will not be marketed.

Soil Solarisation. In warmer climatic areas, solarisation has been effective for disinfesting containerised soil or growing medium and soil in cold frames, as well as soil in open fields. For soil in containers, monitor the temperatures of the growing medium closely to ensure that it is high enough to control pests by placing a soil thermometer into the center of the mass of the soil mix. Planting media can be solarised either in bags or flats covered with transparent plastic or in layers 3- to 9-inches wide sandwiched between two sheets of plastic. A double layer of plastic can increase soil temperature by up to 50°F. In warmer areas of California, soil inside black plastic sleeves can reach 158°F (70°C) during solarisation, equivalent to target temperatures for soil disinfestation by aerated steam.

At this temperature, soil is effectively solarised within 30 minutes. At 140°F (60°C), soil is solarised in 1 hour.

In open fields, soil is more easily covered with a single layer of plastic. Soil temperatures will be lower except at the surface, so plastic should be left in place for 4 to 6 weeks. For maximum effectiveness and treatment predictability, solarise open fields only in warmer climatic areas, unless previous testing has given consistently desirable results. Solarisation is acceptable for organic production.

Heat. Heating the soil is very effective and has the advantage over chemical treatment in that the soil can be planted immediately after cooling. Many plant pathogens are killed by short exposures to high temperatures; however, experience has shown that the soil temperatures should be maintained for approximately 30 minutes. Most plant pathogens can be killed by temperatures of 140°F (60°C) for 30 minutes; however, some viruses (as well as weed seeds) may survive this treatment. (Where weed seeds are a problem, a higher treatment temperature is required, but the tobacco mosaic virus [TMV] and some weed seeds still will not be killed by the higher temperature.)

Although pure steam at sea level is at 212°F (100°C), the temperature at which steam is used to treat soil is usually about 180°F because of air that is present in the steam or in the soil being treated. If air is mixed with steam, the temperature of the steam air mixture can be closely controlled, depending on the ratio of air to steam. It has been demonstrated that some diseases, such as Rhizoctonia damping off, are much less severe in soil that has been treated at 140°F rather than at 180°F. Experience will tell the grower at just what temperature to treat soil. As a starting point try 140°F (60°C) for 30 minutes.

If a cement mixer is used to heat a bulk quantity of soil, generally it is not necessary to introduce air into the steam because a large amount of air is present in the mixer and the temperature can be controlled by simply regulating the flow of steam. Expensive air blowers are not required for this method.

Steam heating of containers filled with soil in vaults likewise may not require the introduction of air into the steam to control the temperature. However, circulation within the vault to insure even distribution of heat is important. Circulating fans can be located within or external to the vault and the steam can be introduced into the recirculating air. Leave space between the vaults and check temperatures throughout the vault to insure that there is good circulation of steam air.

Soil Fumigants. The most useful soil fumigants are methyl bromide and chloropicrin. Although registered for use, metam-sodium (Vapam), and dazomet (Basamid) are not very effective for controlling many soil-borne pathogens, including Verticillium and Fusarium oxysporum. Fumigants such as methyl bromide are a source of volatile organic compounds (VOCs) but are not reactive with other air contaminants that form ozone; methyl bromide depletes ozone. Fumigate only as a last resort when other management strategies have not been successful or are not available.

Methyl bromide is a gas at temperatures over 40°F. It escapes rapidly from soil if not applied under a gas proof cover. Polyethylene sheeting is commonly used to confine methyl bromide, although the gas does slowly escape through polyethylene. Methyl bromide is probably the most versatile of the soil fumigants because of its ability to diffuse rapidly through the soil and kill many kinds of organisms, weeds, and many seeds. Soil generally can be planted in just a few days after removal of plastic covers, although there are exceptions. A few plants such as Allium spp., carnations, and snapdragons are sensitive to and may be damaged by inorganic bromide that remains in the soil following fumigation. Leaching the soil with water before planting is helpful in reducing the amount of bromide in the rooting area. Methyl bromide alone usually does not kill all of the micro-sclerotia of Verticillium dahliae, the fungus that is the causal agent of Verticillium wilt. When this fungus occurs it is often necessary to include chloropicrin in the fumigant.

Methyl bromide may be injected by chisels if the soil is covered immediately by plastic. It is the only soil fumigant that needs to be applied under plastic covers. This is a restricted use material and requires a permit from the county agricultural commissioner to be purchased and/or applied. With the impending loss of methyl bromide as a registered funigant, research is being conducted to find viable alternatives because of ozone depletion concerns.

Chloropicrin (trichloronitromethane) must be injected into soil. It is the best fumigant for controlling Verticillium dahliae. It generally is combined with methyl bromide in various mixtures depending upon the organisms in the soil. If used alone, a water seal may be used to confine the gas; however, the gas is very objectionable and irritating (it is commonly known as tear gas) and, if not effectively confined, it may drift to inhabited areas. This is a restricted use material and requires a permit from the county agricultural commissioner to be purchased and/or applied.

Soil Fungicides. Some fungicides work best if incorporated before planting. Others may be incorporated or applied after sowing or planting. Some soil fungicides control a narrow range of organisms while others control a wide range of organisms. Some of the narrow range chemicals are the most effective in controlling a specific organism. Combinations are used to increase the number of organisms controlled.

Fosetyl-Al (Aliette) is active against Phytophthora species and some Pythium species. It is applied as a soil drench or as a foliar spray but is more effective as a foliar spray. It is absorbed by foliage and moves into roots. It is used as a drench at 0.8 to 1.6 lb a.i./1000 sq ft using 0.5 to 1.5 pt water/sq ft. As a foliar spray it is applied at 2 to 4 lb a.i./100 gal water. The active ingredient in Soil Gard is Gliocladium virens, a soil-borne fungus that under certain conditions helps provide control of Pythium and Rhizoctonia fungi.

Iprodione (Chipco 26019) is used at 0.2 lb a.i./100 gal water applied as a drench (1–2 pt/sq ft) at seeding or transplanting. It is effective against Rhizoctonia damping off, Sclerotinia, and gray mold. Some iprodione is absorbed by plant parts.

Mefenoxam (Subdue Maxx) is active against Pythium, Phytophthora, and downy mildews but is not effective against pathogens outside of this group of organisms. This material replaces the fungicide metalaxyl. It is applied at planting as a drench and periodically thereafter as needed. Mefenoxam is also available in a granular formulation to use before planting. It is water-soluble and readily leached from soil. It is absorbed by plant parts including roots; movement in the plant is primarily in the xylem. Use of this material over a period of time may lead to resistance.

PCNB (Terraclor), also called quintozene, is very active against diseases caused by Rhizoctonia solani and Sclerotinia spp. and is the best available material for southern wilt caused by Sclerotium rolfsii. It is insoluble in water and must be thoroughly mixed with soil to reach its desired depth of control. It works through vapour action and has good residual action. It is inactive against Pythium pathogens. It is used at 0.5 to 1 lb a.i./1000 sq ft and mixed into the top 2 inches of soil for control of Rhizoctonia damping-off. Germination of some seeds may be inhibited and small plants may be stunted by this fungicide.

Streptomyces griseoviridis (Mycostop) is a biofungicide reported to help control seed rot, root and stem rot, and wilt caused by Alternaria and Phomopsis in container-grown ornamentals. In the greenhouse it may suppress Botrytis gray molds and root rots of Pythium, Phytophthora, and Rhizoctonia.

Thiophanate-methyl (Fungo Flo, Cleary's 3336, Zyban, Systec, etc.) is generally applied after sowing. It helps to control Rhizoctonia diseases, cottony rot, Thielaviopsis rots, and some Cylindrocladium diseases. It is not effective against Pythium spp., Phytophthora spp., Sclerotium rolfsii, Botrytis spp., or Fusarium spp. It is used at 0.5 lb a.i. or less/100 gal water and applied as a drench or heavy spray (1-2 pt/sq ft). Thiophanate-methyl is absorbed by plant parts exposed to the chemical. Roots may absorb the fungicide (or its breakdown product carbendazim), which moves in the xylem to transpiring leaves.

Trichoderma spp. (Garden Solutions, Root Guardian) is a biological fungicide reported to help provide control of root diseases caused by Pythium, Rhizoctonia, and Fusarium in nursery and greenhouse crops. It is formulated as a seed protectant, a soil drench, and as granules.

Triflumizole (Terra Guard) is a protectant fungicide used as a cutting soak, soil drench, foliar spray, or through chemigation for control of Cylindrocladium spp. Its use is restricted to enclosed commercial structures such as greenhouses and shade houses.

Seed Treatments. Streptomyces griseoviridis (Mycostop) is used as a seed treatment for damping-off and early root rots for ornamentals planted in fields or greenhouse. Captan and thiram are also seed treatments; they are generally applied at rates of 1 to 4 oz/100 lb seed, but they offer only a small degree of protection.

Materials used for bulb or corm dips include thiabendazole (Mertect), which controls Fusarium basal rot and Penicillium blue mold.

Treatment of Containers and Equipment. Debris, soil, and plant material cling to containers and equipment; thoroughly wash equipment to remove all soil or planting mix particles. Heat treatment is effective in killing the plant pathogens that adhere to containers or that are in the debris. Where steam is not available, hot water or solarisation can be very effective. Most plastic can be treated with hot water at temperatures that cause minimal softening. The minimum water temperature should be 140°F (60°C) whenever possible. Treatment time can be as short as 1 minute. Longer treatment times are more reliable and the container or equipment must reach at least 140°F (60°C). For solarisation, containers should be moistened, stacked, and placed beneath a double-layer tent. Incubate for 30 minutes at or above 158°F (70°C) or 1 hour at or above 140°F (60°C).

Sodium hypochlorite (the active ingredient in bleach) is effective in killing some types of fungal spores and bacteria. It penetrates clinging soil and plant material very poorly. It is effective only as a surface disinfectant, so containers, tools, etc., must be free of soil and plant material and clean before treatment. Sodium hypochlorite is generally used as a surface disinfectant at 0.5 per cent. To achieve this concentration of sodium hypochlorite, household bleach can be diluted 1 part bleach to 10 parts water. For known contaminated materials, a stronger solution diluted 1 part bleach to 5 parts water (1 per cent sodium hypochlorite), may be more effective in killing pathogens. Allow the solution to be in contact with non-porous materials for a minimum of 5 to 10 minutes, then rinse well with clean water to remove bleach and avoid phytotoxicity. Bleach dilutions must be made fresh each day because once diluted, the effectiveness of the solution diminishes over time. Debris, potting mix, and other residues left over in bleach washes will also reduce bleach concentration.

MANAGE SOIL-BORNE PATHOGENS IN VEGETABLES

Farming operations will either help or hinder management of soil-borne pathogens, and, to head off losses, vegetable growers must learn which pathogens they face and tailor practices accordingly, says a University of California, Davis plant pathologist.

Krishna Subbarao, who is based at the USDA Station in Salinas, outlined management principles and pointers at a recent plant disease seminar in Salinas.

He said pathogens that cause disease in vegetable crops either survive in soil or have some cycle in the soil that enables them to infect the crop to cause significant losses in yield and/or quality. They are often difficult to predict, detect, and diagnose. Problems magnify when one or more pathogens occur in a field simultaneously.

They fall into two categories : those that survive in the soil anywhere from a few months to several years and those that are transient in the soil. The first includes, he said, "the common culprits : fungi, nematodes, and viruses."

The second group is made up of various bacteria.

Production methods, including cropping history, influence where pathogens occur in the soil. Vertically, most are found in the top 10 inches of soil, where host roots and tissues and other organic matter are found. Horizontally, pathogens survive on residue of diseased tissue where host plants have grown.

Subbarao said tillage operations break up, move, and bury crop residue with the pathogens that attach to it in the soil. While deeper tillage can spread some of them farther into the soil profile, exposure to heat, cold, and drying closer to the surface may kill others.

Soil type, texture, pH, moisture, temperature, and nutrient levels are among factors that govern how pathogens behave. Well-aerated, well-drained soils create conditions that discourage root diseases, while poorly drained soils tend to favour the survival of several, including pythium, phytophthora, fusarium, and verticillium. Only a few root diseases, such as common scab of potato, are favoured by drier soils.

Soil pH Factor

Soil pH has significant bearing on some soil-borne pathogens. For example, clubroot on crucifers is severe in acidic soils with a pH of 5.7 or less. It has been managed in commercial broccoli fields in the Salinas Valley for years by liming the soil to raise the pH to levels of 5.7 to 6.2. The disease is virtually eliminated with pH levels greater than 7.3 to 7.4.

Some crop residues have been found to be antagonistic to certain soil pathogens.

Plant nutrition and its links to soil pathogens have been the subject of much research, and Subbarao said, in general, high nitrogen levels tend to keep a host's fleshy tissue longer in a vegetative state, making it more vulnerable to attack from pathogens. However, he added, some pathogens are predisposed to invade plants that are weakened by shortages of soil nitrogen.

Nitrogen Type

Not to be overlooked is the type of nitrogen available to the crop and the resulting soil chemistry that favours pathogens. Subbarao and others have reported that the positive charge of an ammonium ion attracts it to plant roots, causing the release of hydrogen ions into the surrounding soil. As the hydrogen ions lower the soil pH, diseases that thrive in acid soil are encouraged. On the other hand, nitrates favour diseases such as vascular wilts by altering the pathogens' virulence and increasing the susceptibility of hosts.

He said diagnosing soil-borne diseases is essentially the same as for those on the plant above ground. He recommended preparing a list of known suspected soil-borne pathogens, which over time can help narrow the group of pathogens to watch.

Complete examination of the entire plant will reveal all symptoms to monitor for a profile of the plant's condition.

Relating symptoms of plants having symptoms to site in the field and irrigation or other cultural practices will help determine disease distribution. Plant samples should be analysed by a qualified laboratory to confirm the true cause of a disorder. Keeping a record of soil-borne diseases found at each site in a field will create a historic record of pathogens. This is important for future reference in selecting crop rotations because many of the pathogens may linger for years on a site. Moving to the management of soil-borne diseases, Subbarao said the severity of a disease is determined by the pathogen's primary inoculum acting with the particular host plant and environmental conditions. The object is to reduce the amount of inoculum needed to set off an infection in the crop. The amount needed for verticillium wilt in strawberries is quite small, just two micro-sclerotia per gram of soil.

Among management options are fumigation, solarisation, host plant resistance, and green manure crops. Success with biological controls has been spotty. Cultural controls in combination with other options may be needed.

Sub-surface Drip

Subbarao said in the case of Sclerotinia minor, the cause of lettuce drop, sub-surface drip irrigation can be an answer. "The more uniform moisture of furrow irrigation encourages Sclerotinia, and the drier soil with drip means less of the pathogen."

He went on to say that his research has shown that if the soil under drip irrigation is kept dry enough about 3 centimeters around and below the plant the disease can be managed. The most familiar soil-borne diseases, including seed decay, damping off of seedlings and root and crown rots, are rots that go to sub-surface tissues or vascular wilts starting with root infections. But according to Subbarao, a few soil-borne pathogens cause foliar disease symptoms and damage that appear on the above ground parts.

One example is lettuce anthracnose, whose micro-sclerotia survive in the soil. When raindrops splash soil laden with these structures onto lettuce leaves, the fungus causes a leaf-spot disease. In like fashion, Sclerotinia sclerotiorum bodies rest in the soil until certain environmental conditions trigger development of mushroom-like apothecia that release aerial spores. The spores then find their way to crop foliage and cause a foliar disease symptom.

Chapter 18

FOOD-BORNE BACTERIAL PATHOGENS

An understanding of the growth characteristics and sources of bacterial pathogens in foods is essential to conducting a hazard analysis of a food and subsequently controlling the identified hazards. On gram-negative rods and gram-positive rods and cocci. The pathogens within each group have some similarities in addition to their gram stain. For example, gram negative rods are non-spore formers and tend to have a fecal source. On the other hand, gram positive rods and cocci can be spore formers and are typically associated with environmental sources like soil and sediments.

GRAM-NEGATIVE RODS

Campylobacter

Campylobacter jejuni infection, called Campylobacteriosis, causes diarrhea, which might be watery or sticky and might contain blood. Other symptoms include fever, abdominal pain, nausea, headache, and muscle pain. Onset of illness occurs 2-5 days after eating contaminated food or water and lasts between 7 and 10 days, with a relapse in 25 per cent of cases. While Campylobacter infections are self-limiting, antibiotics can further limit the amount of time that bacteria are shed in the feces of infected individuals.

The infective dose is considered to be small; human feeding studies suggest that as few as 400-500 bacteria might cause illness. While the pathogenic mechanism is not completely understood, it is an infection. Estimated numbers of cases of campylobacteriosis exceed two to four million per year. In fact, it is considered the leading cause of human diarrheal illness in the U.S. and is reported to cause more disease than Shigella and Salmonella spp. combined. Despite this, death is rare, with one fatality per 1,000 cases. Those most frequently afflicted are children under 5 years and young adults ranging from 15 to 29.

Raw and undercooked chicken, raw and improperly pasteurised milk, raw clams, and non-chlorinated water have been implicated in campylobacteriosis. The organism has been isolated from crabmeat. It is carried by healthy chickens and cows and can be isolated from flies, cats, and puppies.

Campylobacter is unique because of its special oxygen requirements. It is micro-aerophilic, which means it requires reduced levels of oxygen to grow — about 3-15 per cent oxygen (conditions similar to the intestinal tract). Also, it will not grow at temperatures below 86°F or at salt levels above 1.5 per cent. The organismsis considered fragile and sensitive to environmental stresses like drying, heating, disinfection, acid, and air which is 21 per cent oxygen. It requires a high water activity and fairly neutral pH for growth.

The controls include proper cooking and pasteurisation, proper hygienic practices by food handlers to prevent recontamination, and adequate water treatment.

Yersinia — Yersinia spp : Y. entercolitica; Y. pseudotuberculosis; Y. pestis

Of the 11 recognised species of Yersinia, three are known to be pathogenic to humans — enterocolitica, pseudotuberculosis, and pestis. Only enterocolitica and pseudotuberculosis are recognised as food-borne pathogens. Y. pestis, the micro-organism responsible for the black plague, is not transmitted by food and so is not addressed below.

Onset of illness for yersiniosis is about 3 to 7 days, but periods of up to 11 days have been reported. The illness usually lasts 1 to 3 days, but in some cases it might persist for 5 to 14 days or longer. Yersiniosis is often characterised by gastroenteritis with diarrhea and/or vomiting, but fever and abdominal pain are the hallmark symptoms. Yersinia infections mimic appendicitis, which has led to unnecessary appendectomies. Both enterocolitica and pseudotuberculosis have been associated with reactive arthritis, which might occur even in the absence of fever and abdominal pain. Another complication is septicemia, an infection of the blood system. This is rare. Fatalities are also rare.

The infective dose of Yersinia has not been determined. Yersiniosis is rare in U.S.; CDC estimates that only 17,000 cases or so occur annually in the U.S.. Yersiniosis is a far more common problem in Northern Europe, Scandinavia, and Japan. As usual, the most susceptible populations — both for the condition itself and for possible complications — are the very young, the debilitated, the very old, and those undergoing immunosuppressive therapy. Yersinia can be found in raw vegetables, milk, ice cream, cakes, pork, soy products, salads, oysters, clams, and shrimp. They are found in the environment, such as in lakes, streams, soil, and vegetation. They have isolated from the feces of dogs, cats, goats, cattle, chinchillas, mink, and primates; and in the estuarine environment. Many birds, including waterfowl and seagulls, might be carriers.

The food-borne nature of yersiniosis is well established, and numerous outbreaks have occurred worldwide. Two outbreaks in Quebec, Canada, in the mid-1970s affected 138 children and were traced to raw milk. In the U.S., an outbreak in New York in 1976 affected 217 students. In this case, pasteurised chocolate milk was implicated. A 1980 outbreak in Washington affected 87 people. The source of contamination was traced to unchlorinated spring water used in packaging tofu. More outbreaks occurred in 1983 in a tri-state area in the south-east, and again pasteurised milk was implicated.

Yersinia are facultative anaerobes. They are psychrophilic organisms, with a minimum growth temperature of 30°F. Yersinia love cold and can withstand repeated freesing and thawing. Other than that, Yersinia are pretty typical for gram negative bacteria. They have a high water activity and relatively neutral pH requirement, along with a low salt tolerance. Yersinia can be eliminated through pasteurisation or the use of commercial sanitisers.

Yersinia are controlled by proper cooking or pasteurisation, proper food handling to prevent recontamination, adequate water treatment, and preventing time-temperature abuse. Proper use of sanitisers is also an effective control.

Salmonella

There are four syndromes of human salmonellosis — Salmonella gastroenteritis, typhoid fever, non-typhoidal Salmonella septicemia, and asymptomatic carrier. Salmonella gastroenteritis might be caused by any of the Salmonella species other than Salmonella Typhi and is usually a mild, prolonged diarrhea.

True typhoid fever is caused by infection with Salmonella Typhi. While fatality rates might exceed 10 per cent in untreated patients, they are less than 1 per cent in patients who receive proper medical treatment. Survivors might become chronic asymptomatic carriers of Salmonella bacteria. Asymptomatic carriers show no symptoms of illness yet are capable of passing the organisms to others.

Non-typhoidal Salmonella septicemia might result from infection with any of the Salmonella species and can affect virtually all organ systems, sometimes leading to death. Survivors might become chronic asymptomatic carriers of Salmonella bacteria. In this manual the discussion will be limited to Salmonella gastroenteritis because it is the most common form in the U.S.

Although symptoms might appear a few hours after eating contaminated food, it might take one or more days. Symptoms include nausea, vomiting, abdominal cramps, diarrhea, fever, chills, and headache. In some cases, reactive arthritis or Reiter's syndrome can occur up to three weeks after the other symptoms. Depending on the host, the dose, and the strain characteristics, symptoms might last 1-2 days, or longer. For particularly susceptible individuals, the infectious dose might be as few as 15-20 organisms. Salmonella cause between 40,000 and 60,000 reported cases per year, and perhaps as many as 3 million unreported cases. Fatalities range from 1 per cent to 4 per cent. While healthy people can get sick from Salmonella, the population most at risk is the young, the old, and the sick.

Salmonella often live in animals - especially poultry and swine - as well as in a number of environmental sources. Salmonella has been found in water, soil and insects, on food-contact surfaces, and in animal feces. They can also survive in a variety of foods, including raw meats and poultry, dairy products and eggs, fish, shrimp and frog legs, yeast, coconut, sauces and salad dressing, cake mixes, cream-filled desserts and toppings, dried gelatin, peanut butter, orange juice, cocoa and chocolate.

In 1985, a salmonellosis outbreak involving 16,000 confirmed cases in six states was traced to milk from one Chicago dairy, where FDA inspectors found a cross-connection between raw and pasteurised milk. An outbreak involving ice cream was the result of transporting ice cream mix in trucks which had previously hauled raw eggs. An orange juice outbreak proved to be an example of Salmonella adapting to an acidic environment.

In 1985, S. enteritidis was blamed for at least 71 illnesses in Maryland, and scrambled eggs from a breakfast bar were implicated. CDC estimates that 75 per cent of S. enteritidis outbreaks are associated with eating raw or inadequately cooked eggs. The U.S. DA published regulations in 1990 establishing a mandatory testing programme for egg-producing breeder flocks and commercial flocks implicated in causing human illnesses. This testing should lead to a reduction in cases of gastroenteritis caused by eating eggs.

Salmonella spp. are also mesophilic organisms which grow best at moderate temperatures and pH, and under conditions of low salt and of high water activity. They are killed rapidly by moderate heat treatment, yet mild heat treatment might enable them to develop heat resistance, up to 185°F. Similarly, they can adapt to an acidic environment, as in the case of the orange juice outbreak mentioned earlier.

Ordinary household cooking, personal hygiene to prevent recontamination of cooked food, and control of time and temperature are generally adequate to prevent salmonellosis.

Shigella

There are four species of Shigella. Because there is little difference in their behaviour, they will be discussed as a group.

The onset time for shigellosis can range from 12-96 hours. Typical symptoms include fever, cramps, tenesmus, inflammation and ulceration of the intestine, and diarrhea. Sometimes the diarrhea can lead to dysentery, which is not usually a life-threatening illness. In malnourished children, immuno-compromised individuals and older adults, the disease might be lethal. Ordinary shigellosis might also be severe, but it is self-limiting. If left untreated, it might last 1 to 2 weeks. The infectious dose might be as few as 10 cells, depending on the age and condition of the host, and the disease is easily transmitted from person to person, with a high secondary attack rate. Approximately 14,000 cases of shigellosis are reported every year, with an estimated total of 300,000 cases. Of the 14,000, an average of four will result in death.

The only significant reservoir for Shigella is humans. Foods associated with shigellosis include salads (potato, tuna, shrimp, macaroni, and chicken), raw vegetables, milk and dairy products, poultry, fruits, bakery products, hamburger, and finfish.

In 1985, a large outbreak of food-borne shigellosis occurred in Texas, involving as many as 5,000 persons. The implicated food was chopped, bagged lettuce, prepared in a central location for a Mexican restaurant chain. A number of outbreaks of shigellosis occurred on college campuses throughout the U.S.. Fresh vegetables from the salad bar were usually implicated, and an ill foodservice worker proved to be the cause.

The growth conditions for Shigella, which are mesophilic organisms, are similar to those of Salmonella. Shigella can survive under various environmental conditions, including low-acid conditions.

Shigella can spread rapidly under the crowded and unsanitary conditions often found in such places as summer camps, refugee camps, and camps for migrant workers, and at mass gatherings, such as music festivals. The primary reasons for the spread of Shigella in foods are poor personal hygiene on the part of food handlers, and the use of improper holding temperatures for contaminated foods. The best preventive measures would be good personal hygiene and health education. Chlorination of water and sanitary disposal of sewage would prevent waterborne outbreaks of shigellosis.

Shigella are closely genetically related to E. coli and could be considered the same species. But, because all strains of Shigella cause severe diarrhea, while most strains of E. coli do not, microbiologists differentiate between the two.

Escherichia Coli

E. coli have long been used as indicator organisms because E. coli are present in the normal gut flora of warm-blooded animals. There are four classes of pathogenic

E. coli – enteropathogenic (EPEC), enterotoxigenic (ETEC), enteroinvasive (EIEC), and enterohemorrhagic (EHEC). All four types have been associated with food-borne disease. Although the growth requirements are similar for each class, the diseases differ, so we will cover each separately.

EPEC. The onset time of EPEC is 17-72 hours, and the disease can last anywhere from six hours to three days. Outbreaks most often affect infants, especially those who are bottle-fed, suggesting that contaminated water is often used to rehydrate infant formulae in underdeveloped countries. Occasionally diarrhea in infants is prolonged, leading to dehydration, electrolyte imbalance and death. Diarrhea can be either watery or bloody. A 50 per cent mortality rate has been reported in third-world countries.

EPEC is highly infectious for infants, and the infectious dose is presumably very low. In the few documented cases of adult disease, the infective dose was greater than 1,000,000 total. Symptomatic and asymptomatic human carriers are believed to be a principle reservoir of EPEC. Many outbreaks of infantile diarrhea due to EPEC were reported in the 1950s, but these have largely disappeared, due to improvements in sanitation and hygiene. EPEC has been implicated in day-care and nursery outbreaks, and some cases of travellers' diarrhea. Examples of outbreaks affecting adults include an outbreak in Sweden where EPEC was isolated from both the water supply and the feces of ill individuals, an outbreak in Britain linked to cold pork, and another outbreak in Britain related to meat pie. Immunity might explain the rare occurrence of EPEC illness in adults.

ETEC. ETEC is perhaps the most widely known. It is also commonly called traveller's diarrhea. ETEC frequently causes diarrhea in infants in less developed countries, and in visitors from industrialised countries. The most frequent symptoms include watery diarrhea, abdominal cramps, low-grade fever, nausea, vomiting, and malaise. Volunteer feeding studies show that a relatively large dose – 100 million to 10 billion bacteria – is probably necessary for infection. The onset of symptoms ranges between 8 and 44 hours. With a high dose, diarrhea can be induced within 24 hours. Infants might require fewer organisms for infection. The disease can last anywhere from 3-19 days, but is usually self-limiting, in infants or debilitated elderly persons, electrolyte replacement therapy might be necessary.

Contamination of food or water does occasionally lead to outbreaks of ETEC. In 1975, more than 2000 staff members and visitors at a park in Oregon developed gastroenteritis caused by ETEC. The source was the park's water supply, which had been contaminated by raw sewage. In 1983, a multi-state outbreak of ETEC gastroenteritis with 169 reported cases was associated with imported French Brie cheese. As is also the case with EPEC, human carriers are believed to be a principle reservoir.

EIEC. EIEC resembles Shigella in many ways. Like Shigella, EIEC produces in humans an invasive, dysenteric form of diarrhea known as bacillary dysentery. The infectious dose of EIEC is believed to be as few as 10 cells.

Dysentery caused by EIEC usually occurs within 12-72 hours following eating contaminated food. This illness is characterised by abdominal cramps, diarrhea, vomiting, fever, chills, and a generalised malaise. Dysentery caused by this organism is generally self-limiting, with no known complications.

EIEC is associated with water, cheese, potato salad, and canned salmon. Again, human carriers are believed to be the reservoir. A major food-borne outbreak in the United States, involving 387 individuals, was traced to imported French Brie and Camembert cheese; it appeared that a water filtering system was malfunctioning at the time the cheese was produced. A second major outbreak in the U.S. occurred on a cruise ship – potato salad served at a cold buffet was implicated. The contamination might have occurred through preparation by an infected food handler or through the use of contaminated raw ingredients.

EHEC. Judging solely by medical records, one might assume that hemorrhagic colitis, caused by EHEC, is not particularly common, but this is probably inaccurate. In the Pacific Northwest, one strain – E. coli O157 : H7 – is thought to be second only to Salmonella as a cause of bacterial diarrhea. Because of the unmistakable symptoms of profuse, visible blood in severe cases, victims are more likely to seek medical attention, but less severe cases are probably more numerous and less likely to be reported.

Onset of the disease is anywhere from 3-9 days. The infective dose is unknown, but is suspected to be similar to that of Shigella (10 organisms). The illness is characterised by severe cramping (abdominal pain) and diarrhea, which is initially watery but becomes grossly bloody. Occasionally vomiting occurs. Fever is either low-grade or absent. The illness is usually self-limited and lasts for an average of eight days. Some individuals exhibit watery diarrhea only.

All people are believed to be susceptible to hemorrhagic colitis. Some victims, particularly the very young, have developed hemolytic uremic syndrome (H.U.S.), characterised by renal failure and hemolytic anemia. The disease can lead to permanent loss of kidney function. In older adults, this illness can have a mortality rate as high as 50 per cent.

The intestinal tract of cattle and other food animals are reservoirs for EHEC. EHEC has been associated with ground beef, raw milk, fermented sausage, apple cider, unpasteurised apple juice, mayonnaise, water, raw vegetables, and club sandwiches. In 1986, an outbreak of hemorrhagic colitis occurred in Washington. Thirty-seven (37) people, ages 11 months to 78 years, developed diarrhea traced to E. coli O157 : H7. Of the 17 patients hospitalised, two died. Ground beef was implicated.

In 1993, a similar outbreak involving undercooked hamburger occurred in Washington, resulting in 447 illnesses and three deaths. During the same year, several other outbreaks along the Pacific Coast were reported; these involved salad bars, raw milk, a church dinner, and a Mexican fast food chain. The most recent outbreak again occurred in Washington in October 1996. The product implicated was unpasteurised apple juice; 70 people became ill, and there was one death.

The growth and survival conditions and controls are similar for all four classes of E. coli so these conditions will be discussed collectively. E. coli are mesophilic organisms. They grow best at moderate temperatures, at moderate pH, and in conditions of high water activity. It has, however, been shown that some E. coli strains are very tolerant of acidic environments and freesing.

Food might be contaminated by infected food handlers who practice poor personal hygiene or by contact with water contaminated by human sewage. Control measures to prevent food poisoning, therefore, include educating food workers on safe food handling techniques and proper personal hygiene, properly heating foods, and holding foods under appropriate temperature controls. Additionally, untreated human sewage should never be used to fertilize vegetables and crops used for human consumption, nor should unchlorinated water be used for cleaning food-contact surfaces.

Prevention of fecal contamination during the slaughter and processing of foods of animal origin is paramount to control food-borne infection of EHEC. Foods of animal origin should be heated sufficiently to kill the organism. Consumers should avoid eating raw or partially cooked meats and poultry and drinking unpasteurised milk or fruit juices.

Vibrios

There are many species of Vibrios, but only four will be covered — Vibrio parahaemolyticus; Vibrio cholerae 01; Vibrio cholerae non-01; and Vibrio vulnificus.

Vibrio parahaemolyticus. V. parahaemolyticus is naturally occurring in estuaries and other coastal waters. Implicated foods are fish and shellfish that are raw, improperly cooked or recontaminated after cooking.

The symptoms associated with Vibrio parahaemolyticus are diarrhea, abdominal cramps, nausea, vomiting, and fever. Illness is usually mild or moderate, with onset in 4-96 hours, and it lasts for 2.5 days. It can also result in septicemia. The infective dose is one million organisms.

Vibrio cholerae 01. Epidemic cholera is caused by Vibrio cholerae 01. Onset ranges from 6 hours to 5 days, and symptoms vary from mild, watery diarrhea to acute diarrhea with characteristic rice water stools. Illness can include abdominal cramps, nausea, vomiting, dehydration and shock, while severe fluid and electrolyte loss might result in death. The infective dose — determined by healthy human volunteer feeding studies — is one million organisms. Poor sanitation and contaminated water supplies spread the disease. Proper sewage treatment is responsible for the near-eradication of epidemic cholera in the U.S. At present, about 20 cases are reported per year, and these are usually the result of travellers returning from developing countries. The organism is found in sewage-contaminated water, and has been associated with various feces-contaminated foods and beverages, including seafood.

Vibrio cholerae non-01. V. cholerae non-01 is generally a less severe gastroenteritis than that caused by V. cholerae 01. The symptoms are diarrhea, abdominal cramps, fever, nausea, and vomiting. Onset of symptoms occurs between 6 hours and 3 days, and these symptoms can last from 6-7 days. This organism can also cause septicemia. As with V. parahaemolyticus, the reservoir for this organism is estuarine water. Illness is associated with raw oysters, but the bacterium has also been found in crabs.

Vibrio vulnificus. V. vulnificus is an estuarine species that can cause wound infections, gastroenteritis, or primary septicemia. It is one of the most severe food-borne infectious diseases, with a fatality rate of 50 per cent for those who contract septicemia. Healthy individuals are most often susceptible to gastroenteritis, while high-risk individuals (those with cirrhosis or other liver disease, diabetes, leukemia or immunosuppression) are particularly susceptible to primary septicemia and so should not eat raw shellfish. The onset of fever, chills, and nausea can occur within 24-48 hours; where septicemia develops, death has been reported to occur within 36 hours.

There have been no major outbreaks — only sporadic individual cases, usually during the warm weather months. Again, as with V. parahaemolyticus, this organism occurs naturally in estuarine waters. So far, only oysters from the Gulf of Mexico have been implicated in illness, but the organism itself has been isolated from both the Atlantic and Pacific Oceans.

Like the other gram negatives, Vibrios are mesophilic and require relatively warm temperatures, high water activity and neutral pH for growth; unlike the others, they also require some salt for growth, and are quite salt-tolerant. They are, however, easily eliminated by a mild heat treatment.

All the Vibrios can be controlled by thorough cooking and the prevention of cross-contamination afterwards. Proper refrigeration prevents proliferation, which is particularly important because of the short generation times for these species. The high infective dose of V. parahaemolyticus makes such control even more vital. To guard against cholerae, food handlers should know the source of their product and be cautious about importing from countries experiencing an epidemic.

COMMON FOOD-BORNE PATHOGENS

Even though the United States has one of the safest food supplies in the world, there are still millions of cases of food-borne illness each year. Here are common food-borne pathogens (disease-causing micro-organisms) with research-based information that includes :

- Cause of illness
- Incubation period
- Symptoms
- Possible contaminants
- Steps for prevention

Bacillus Cereus

- *Cause of Illness :* Large molecular weight protein (diarrheal type) or highly heat-stable toxin (emetic type)
- *Incubation Period :* 30 minutes to 15 hours
- *Symptoms :* Diarrhea, abdominal cramps, nausea, and vomiting (emetic type)
- *Possible Contaminants :* Meats, milk, vegetables, fish, rice, potatoes, pasta, and cheese
- *Steps for Prevention :* Pay careful attention to food preparation and cooking guidelines.

Campylobacter Jejuni

- *Cause of Illness :* Infection, even with low numbers
- *Incubation Period :* One to seven days

- *Symptoms :* Nausea, abdominal cramps, diarrhea, headache — varying in severity
- *Possible Contaminant :* Raw milk, eggs, poultry, raw beef, cake icing, water
- *Steps for Prevention :* Pasteurise milk; cook foods properly; prevent cross-contamination.

Clostridium Botulinum

- *Cause of Illness :* Toxin produced by Clostridium botulinum
- *Incubation Period :* 12 to 36 hours
- *Symptoms :* Nausea, vomiting, diarrhea, fatigue, headache, dry mouth, double vision, muscle paralysis, respiratory failure
- *Possible Contaminant :* Low-acid canned foods, meats, sausage, fish
- *Steps for Prevention :* Properly can foods following recommended procedures; cook foods properly.

Clostridium Perfringens

- *Cause of Illness :* Undercooked meats and gravies
- *Incubation Period :* 8 to 22 hours
- *Symptoms :* Abdominal cramps and diarrhea, some include dehydration
- *Possible Contaminants :* Meats and gravies
- *Steps for Prevention :* Proper attention to cooking temperatures.

Cryptosporidium Parvum

- *Cause of Illness :* Drinking contaminated water; eating raw or undercooked food; putting something in the mouth that has been contaminated with the stool of an infected person or animal; direct contact with the droppings of infected animals.
- *Incubation Period :* Two to 10 days
- *Symptoms :* Watery diarrhea accompanied by mild stomach cramping, nausea, loss of appetite. Symptoms may last 10 to 15 days.
- *Possible Contaminants :* Contaminated water or milk, person-to-person transmission (especially in child daycare settings). Contaminated food can also cause infections.
- *Steps for Prevention :* Avoid water or food that may be contaminated; wash hands after using the toilet and before handling food. If you work in a child care center where you change diapers, be sure to wash hands thoroughly with soap and warm water after every diaper change, even if you wear gloves. During communitywide outbreaks caused by contaminated drinking water, boil drinking water for 1 minute to kill the Cryptosporidium parasite. Allow water to cool before drinking it.

Escherichia coli 0157 :H7

- *Cause of Illness :* Strain of enteropathic E.coli
- *Incubation Period :* Two to four days
- *Symptoms :* Hemorrhagic colitis, possibly hemolytic uremic syndrome
- *Possible Contaminant :* Ground beef, raw milk
- *Steps for Prevention :* Thoroughly cook meat; no cross-contamination.

Giardia Lamblia

- *Cause of Illness :* Strain of Giardia lamblia.
- *Incubation Period :* One to two weeks.
- *Symptoms :* Infection of the small intestine, diarrhea, loose or watery stool, stomach cramps, and lactose intolerance.
- *Possible Contaminant :* Giardia is found in soil, food, water, or surfaces that have been contaminated with the feces from infected humans or animals.
- *Steps for Prevention :* Avoid swallowing contaminated recreational water (pools, hot tubs, fountains, lakes, rivers, ponds) or contaminated bathroom fixtures, toys, changing tables, diaper pails; avoid eating uncooked contaminated food; boil water for 1 minute before use or use a water filter with an absolute pore size of at least 1 micron or rated for "cyst removal." Cholorination or iodination of water may be less effective. Avoid fecal exposure during sexual activity.

Hepatitis A

- *Cause of Illness :* Hepatitis A Virus (HAV)
- Incubation period :
- *Symptoms :* Fever, malaise, nausea, abdominal discomfort
- *Possible contaminants :* Water, fruits, vegetables, iced drinks, shellfish, and salads
- *Steps for prevention :* Carefully wash hands with soap and water after using a restroom, changing a diaper, and before preparing food.

Listeria Monocytogenes

- *Cause of Illness :* Infection with Listeria monocytogenes
- *Incubation Period :* Two days to three weeks
- *Symptoms :* Meningitis, sepsticemia, miscarriage
- *Possible Contaminant :* Vegetables, milk, cheese, meat, seafood
- *Steps for Prevention :* Purchase pasteurised dairy products; cook foods properly; no cross-contamination; use sanitary practices.

Norwalk, Norwalk-like, or Norovirus

- *Cause of Illness :* Infection with Norwalk virus
- *Incubation Period :* Between 12 and 48 hours (average, 36 hours); duration, 12-60 hours
- *Symptoms :* Nausea, vomiting, diarrhea and abdominal cramps
- *Possible Contaminant :* Raw oysters/shellfish, water and ice, salads, frosting, person-to-person contact
- *Steps for Prevention :* Adequate and proper treatment and disposal of sewage, appropriate chlorination of water, restriction of infected food handlers from working with food until they no longer shed virus.

Salmonellosis

- *Cause of Illness :* Infection with Salmonella species
- *Incubation Period :* 12 to 24 hours
- *Symptoms :* Nausea, diarrhea, abdominal pain, fever, headache, chills, prostration
- *Possible Contaminant :* Meat, poultry, egg or milk products
- *Steps for Prevention :* Cook thoroughly; avoid cross-contamination; use sanitary practices.

Staphylococcus

- *Cause of Illness :* Toxin produced by certain strains of Staphylococcus aureus
- *Incubation Period :* One to six hours
- *Symptoms :* Severe vomiting, diarrhea, abdominal cramping
- *Possible Contaminant :* Custard- or cream-filled baked goods, ham, tongue, poultry, dressing, gravy, eggs, potato salad, cream sauces, sandwich fillings
- *Steps for Prevention :* Refrigerate foods; use sanitary practices.

Shigella

- *Cause of Illness :* Water contaminated with human feces and unsanitary food handling
- *Incubation period :* 12 to 50 hours
- *Symptoms :* Abdominal pain, cramps, diarrhea, fever, vomiting, blood, and pus
- *Possible Contaminants :* Salads, raw vegetables, dairy products, and poultry
- *Steps for Prevention :* Practice proper washing and sanitising techniques.

Toxoplasma Gondii

- *Cause of Illness :* Parasitic infection
- *Incubation Period :* Five to 23 days after exposure
- *Symptoms :* In healthy children and adults, toxoplasmosis may cause no symptoms at all, or may cause a mild illness (swollen lymph glands, fever, headache, and muscle aches).

Toxoplasmosis is a very severe infection for unborn babies and for people with immune system problems.

- *Possible Contaminant :* Cat, rodent or bird feces, raw or undercooked food.
- *Steps for Prevention :* Wash hands thoroughly after working with soil, cleaning litter boxes, before and after handling foods, and before eating. Cover sandboxes when not in use.

Vibrio

- *Cause of Illness :* Excretion of toxin from infected fish and shellfish
- *Incubation Period :* Four hours to four days
- *Symptoms :* Diarrhea, abdominal cramps, nausea, vomiting, headache, fever, and chills
- *Possible Contaminants :* Fish and shellfish
- *Steps for Prevention :* Cook fish and shellfish thoroughly.

Yersiniosis

- *Cause of Illness :* Infection with Yersinia enterocolitica
- *Incubation Period :* One to three days
- *Symptoms :* Enterocolitis, may mimic acute appendicitis
- *Possible Contaminant :* Raw milk, chocolate milk, water, pork, other raw meats
- *Steps for Prevention :* Pasteurise milk; cook foods properly; no cross-contamination; use sanitary practices. Retail/Institutional Food Service Food Safety and Management.

GRAM POSITIVE RODS AND COCCI

Bacillus Cereus

Bacillus cereus is a gram-positive, aerobic spore former that causes an intoxication. Two types of toxins can be produced — one results in diarrheal syndrome and the other in the emetic syndrome. Onset for the diarrheal syndrome is 6-15 hours after ingestion, with a duration of 24 hours. The primary symptom is diarrhea; vomiting is rare. Onset for the emetic syndrome is earlier — 30 minutes to

6 hours after eating. As with the diarrheal syndrome, the duration is 24 hours. Food counts of B. cereus greater than 106/gram (1,000,000) indicate active growth and a potential hazard to health.

B. cereus is widely distributed throughout the environment. It has been isolated from a variety of foods, including meats, dairy products, vegetables, fish, and rice. The bacteria can also be found in starchy foods such as potato, pasta, and cheese products, and in food mixtures, such as sauces, puddings, soups, casseroles, pastries, and salads.

Fried rice is a leading cause of B. cereus emetic-type food poisoning in the U.S. The organism is frequently present in uncooked rice, and its heat-resistant spores survive cooking. If the rice is then held at room temperature, the spores might germinate and multiply. The toxin produced can survive heating (for instance, stir frying), and many people are unaware that cooked rice is a potentially hazardous food.

This organism will grow at temperatures as low as 39°F, at a pH as low as 4.3, and at salt concentrations as high as 18 per cent. Unlike other pathogens, it is an aerobe, and will grow only in the presence of oxygen. Both the spores and the emetic toxin are heat-resistant.

Listeria Monocytogenes

Unlike B. cereus, Listeria monocytogenes is not easily controlled by refrigeration. Listeriosis, the disease caused by this organism, can produce mild flu-like symptoms in healthy individuals. In susceptible individuals, including pregnant women, newborns, and the immuno-compromised, the organism might enter the blood stream, resulting in septicemia. Ultimately, listeriosis can result in meningitis, encephalitis, spontaneous abortion, and still-birth. The onset of disease might range from a few days to three weeks. The infectious dose is unknown.

L. monocytogenes can be isolated from soil, silage, and other environmental sources. It can also be found in man-made environments, such as food processing establishments. Generally, the drier the environment, the less likely the environmental will harbour this organism.

L. monocytogenes has been associated with raw or inadequately pasteurised milk, cheeses (especially soft-ripened types), ice cream, raw vegetables, fermented sausages, raw and cooked poultry, raw meats, and raw and smoked fish. In 1985, Mexican-style cheese led to at least 46 still-births in California. The consumption of large quantities of smoked mussels in New Zealand is reported to have caused two women to experience spontaneous abortions. The CDC has linked listeriosis with eating raw hot dogs or undercooked chicken.

L. monocytogenes is a psychotropic facultative anaerobe. It can survive some degree of thermal processing, but can be destroyed by cooking to an internal temperature of 158°F for two minutes. It can also grow at refrigerated temperatures below 31°F. Reportedly, it has a doubling time of 1.5 days at 40°F.

There is nothing unusual about Listeria's pH and water activity range for growth.

L. monocytogenes is salt-tolerant : it can grow in up to 10 per cent salt and has been known to survive in 30 per cent salt. It is also nitrite-tolerant.

Prevention of recontamination after cooking is a necessary control; even if the product has received thermal processing adequate to inactivate L. monocytogenes, the widespread nature of the organism provides the opportunity for recontamination. Furthermore, if the heat treatment has destroyed the competing microflora, L. monocytogenes might find itself in a suitable environment without competition.

Clostridium Perfringens

Clostridium perfringens is an anaerobic spore former and is a common cause of food-borne gastroenteritis. Perfringens poisoning is characterised by intense abdominal cramps and diarrhea, which begin 8-22 hours after eating contaminated food. The food must contain large numbers (100,000,000 or more) of the bacteria in order to produce toxin in the intestine. The illness is usually over within 24 hours, but less severe symptoms might persist in some individuals for 1-2 weeks. A few deaths have been reported as a result of dehydration and other complications.

CDC estimates that there are 10,000 cases per year. Of these, approximately 1,200 are reported. The large number of cases and the small number of outbreaks are jointly attributable to institutional feeding, such as school cafeterias and nursing homes. Perfringens poisoning most frequently occurs in the young and older adults.

C. perfringens is widely distributed in the environment and is frequently found in the intestines of humans and many domestic and feral animals. Spores of the organism persist in soil and sediments. C. perfringens has been found in beef, pork, lamb, chicken, turkey, stews, casseroles and gravy. In one 1984 outbreak involving 77 prison inmates, the implicated food was roast beef. Soon afterwards, there was a second outbreak which involved many of the same people, and on that occasion, the food implicated was ham. The cause in these instances was determined to have been inadequate refrigeration and insufficient reheating of the implicated foods.

In 1985, a large outbreak of C. perfringens gastroenteritis occurred among factory workers in Connecticut, and some 600 employees were affected. In that case, gravy that was prepared 12-24 hours before serving and inadequately cooled was implicated.

Clostridium perfringens is a mesophilic organism. Because it is also a spore-former, it is quite resistant to heat, and temperatures for growth range from 50°F to 125°F. The pH, water activity and salt ranges for growth are fairly typical.

Cooking the spores does not kill them. Cooking encourages them to germinate when the food reaches suitable temperature. Rapid, uniform cooling after cooking is critical. In virtually all outbreaks, the principal cause of perfringens poisoning is failure to properly refrigerate previously cooked foods, especially when it is prepared in large portions. Proper hot-holding (above 135°F) and adequate reheating of cooked, chilled foods (to a minimum internal temperature of 75°F) are also necessary controls. Educating food handlers remains a critical aspect of control.

C. Botulinum

Like perfringens, C. botulinum is an anaerobic spore-former. There are seven types of C. botulinum — A, B, C, D, E, F, and G — but the types that will discussed be are type A, which represents a group called proteolytic botulinum, and type E, which represents the non-proteolytic group. The reason for the distinction is the proteolytic organisms' ability to breakdown protein.

This organism is one of the most lethal food-borne pathogens. The infectious dose is exceedingly low; a few nanograms of toxin can cause illness, and everyone is susceptible. Typically, the onset might be from 18-36 hours after eating contaminated food, but this can vary from 4 hours to 8 days. Symptoms include weakness and vertigo, followed by double vision and progressive difficulty in speaking, breathing, and swallowing. There might also be abdominal distention and constipation. The toxin eventually causes paralysis, which progresses symmetrically downward, starting with the eyes and face, and proceeding to the throat, chest, and extremities. When the diaphragm and chest muscles become involved, respiration is inhibited, and death from asphyxia results. Treatment includes early administration of antitoxin and mechanical breathing assistance. Mortality is high; without antitoxin, death is almost certain.

There is a variation of botulism known as infant botulism. In this case, the toxin is formed in the intestinal tract rather than preformed in the food. Honey is the only food that has been definitely linked to this disease, and it has occurred only in infants. The symptoms begin with constipation, followed by loss of appetite, lethargy, general weakness, pooled oral secretions, altered cry, and loss of head control, which is striking. Given the potential danger, one should never feed an infant honey.

C. botulinum is widely distributed in nature and can be found in soils, sediments from streams, lakes and coastal waters, the intestinal tracts of fish and mammals, and the gills and viscera of crabs and other shellfish. Type E is most prevalent in fresh water and marine environments, while type A is generally found terrestrially.

C. botulinum has been a problem in a wide variety of foods — canned foods, acidified foods, smoked and uneviscerated fish, stuffed eggplant, garlic-in-oil, baked potatoes, sauteed onions, black bean dip, meat products, and marscapone cheese.

Two outbreaks in the 1960's involved vacuum-packaged fish (smoked ciscos and smoked chubs). The causative agent in each case was C. botulinum type E. The food was packed without nitrites, with low levels of salt, and were temperature-abused during distribution, all of which contributed to the formation of the toxin.

There were no obvious signs of spoilage because aerobic spoilage micro-organisms were inhibited by the vacuum packaging, and because type E does not produce any offensive odors.

In 1987, there were eight cases in which kapchunka – an uneviscerated, salted, air-dried whitefish – was implicated. It was believed that the fish contained low levels of salt during air drying at room temperature, which allowed for the toxin formation. The outbreak resulted in one death. Three cases of botulism in New York were traced to chopped garlic bottled in oil, which had been held at room temperature for several months before it was opened. Presumably, the oil created an anaerobic environment.

Type A and type E vary in their growth requirements. Minimum growth temperature for type A is 50°F, while type E will tolerate conditions down to 38°F. Type A's minimum water activity is 0.94, and type E's is -.97 – a small difference but important when controlling the organism. The acid-tolerance of type A is reached at a pH of 4.6, while type E can grow at a pH of 5. And type A is more salt-tolerant; it can handle up to 10 per cent, while 5 per cent is sufficient to stop the growth of type E.

Although the vegetative cells are susceptible to heat, the spores are heat-resistant and able to survive many adverse environmental conditions. Type A and type E differ in the heat-resistance of their spores. Compared to type E, type A's resistance is relatively high. By contrast, the neuro-toxin produced by C. botulinum is not resistant to heat, and can be inactivated by heating for 10 minutes at 176°F.

There are two primary strategies to control C. botulinum. The first is destruction of the spores by heat (thermal processing). The second is to alter the food to inhibit toxin production – something that can be achieved by acidification, controlling water activity, adding salt and/or preservatives, and refrigeration. Water activity, salt, and pH can each be individually considered a full barrier to growth, but very often these single barriers – a pH of 4.6 or 10 per cent salt – are not used because they result in a product which is unacceptable to consumers. For this reason, multiple barriers are used.

One example of a product using multiple barriers is pasteurised crabmeat stored under refrigeration. Type E is destroyed by the pasteurisation process, while type A is controlled by refrigerated storage.

Another example of multiple barriers is hot-smoked, vacuum-packed fish. Vacuum-packaging provides the anaerobic environment necessary for the growth of C. botulinum, even as it inhibits the normal aerobic spoilage flora that would otherwise offer competition and exhibit signs of spoilage. So heat is used to weaken the spores of type E, which are then further controlled by the use of salt, sometimes in combination with nitrites. Finally, the spores of type A are controlled by refrigeration.

Vacuum-packaging of foods that are minimally processed, like sous vide foods, allows the survival of C. botulinum spores while wiping out competing microflora. If no control barriers are present, the C. botulinum might grow and produce toxin, particularly if there is temperature abuse. Given the frequency of temperature abuse documented at the retail and consumer levels, this process is safe only if temperatures are carefully controlled to below 38°F throughout distribution.

Vacuum-packing is also used to extend the shelf-life of the product. Because this provides additional time for toxin development, such food must be considered a high risk. Controls can be used to prevent the recurrence of such incidents as the 1987 outbreak caused by uneviserated fish. Any seafood product which will be preserved using salt, drying, pickling, or fermentation must be eviscerated prior to processing; the only exception is small fish (less than five inches in length), which will instead be processed to inhibit the formation of C. botulinum toxin – something that can be done by maintaining a water phase salt of 10 per cent, a water activity of below 0.85, or a pH of 4.6 or less.

Staphylococcus. Aureus

Staphylococcus. aureus is a gram-positive cocci that grows in irregular clusters and produces a highly heat-stable toxin. Staphylococcal food poisoning is one of the most economically important food-borne diseases in the U.S., costing approximately $1.5 billion each year in medical expenses and loss of productivity. Onset is rapid, usually within four hours of ingestion, and the most common symptoms are nausea, vomiting, abdominal cramps, diarrhea, and prostration. Recovery usually takes two days, but can take longer in severe cases. S. aureus food poisoning is usually considered a mild, self-limiting illness with a low mortality rate; however, death has been known to occur among infants, older adults, and severely debilitated individuals. The infective dose is less than 1.0 microgram of toxin, and this toxin level is reached when the S. aureus population reaches 100,000 cells per gram in the food.

S. aureus can be found in air, dust, sewage, and water, although humans and animals are the primary reservoirs. S. aureus is present in and on the nasal passages, throats, hair and skin of at least one out of two healthy individuals. Food handlers are the main source of contamination, but food equipment and the environment itself can also be sources of the organism.

Foods associated with S. aureus include poultry, meat, salads, bakery products, sandwiches, and dairy products. Due to poor hygiene and temperature abuse, a number of outbreaks have been associated with cream-filled pastries and salads such as egg, chicken, tuna, potato, and macaroni.

S. aureus grows and produces toxin at the lowest water activity (0.85) of any food pathogen. Like type A botulinum and Listeria, S. aureus is salt-tolerant and will produce toxin at 10 per cent.

Foods that require considerable handling during preparation and that are kept at slightly elevated temperatures after preparation are frequently involved in staphylococcal food poisoning. And, while S. aureus does not compete well with the bacteria normally found in raw foods, it will grow both in cooked food and in salted food where the salt inhibits spoilage bacteria. Because S. aureus is a facultative anaerobe, reduced oxygen packaging can also give it a competitive advantage.

The best way to control S. aureus is to ensure proper employee hygiene and to minimise exposure to uncontrolled temperatures. While the organism can be killed by heat, the toxin cannot be destroyed even by thermal processing.

FOOD-BORNE BACTERIA

Harmful bacteria are the most common causes of food borne illnesses. Most of the bacteria that cause these illnesses go undetected because they have no odour and cause no change in the colour or the texture of the food that they are growing on. Therefore, it would be fairly easy to cause an epidemic in a fairly large population using a bacteria to cause infection. Bioterrorism is a constant threat in the U.S. today and must be monitored in every way possible. Food is another way that bioterrorists could effect America. By infecting major foods consumed in the U.S. everyday an outbreak of illness could be implicated without much trouble. There are many bacteria that could be used to cause these illnesses, and some of the more effective ones are listed below telling what they are and the symptoms that their diseases entail. Food should always be handled very carefully even when not considering it from a bioterrorist point of view. Bacteria from food can come from many different places including : raw or undercooked meat, unwashed or contaminated produce, seafood, food handling, and cross-contamination from other areas. Cross-contamination can happen wherever the food was processed or packaged or in your own home when countertops and other areas where food was handled is not properly cleaned.

There are many preventative measures that can be taken to ensure not to ingest pathogenic bacteria and some of these are listed below :

- Make sure that food is promptly refrigerated. If prepared food is left in room temperature for more than two hours it may not be safe to eat. Make sure that your refrigerator is set at 40°F or lower and your freezer at 0°F or lower.
- Make sure that food is cooked at the appropriate temperature and for long enough or all of the bacteria will not be eliminated. (145°F for red meat, 160°F for pork and ground beef, 165°F for ground poultry and 180°F for whole poultry)
- Prevent cross-contamination. Keep raw meat, poultry, seafood, and their juices away from other foods that are ready to eat.
- Ensure that food in handled properly by washing hands before and after preparation. Also make sure that surfaces where food in handled is always clean.
- Never defrost food on the kitchen counter.
- Do not pack the refrigerator. Cool air must circulate to keep food safe.

Everyone is always at risk when food borne bacteria are concerned so food must always be handled carefully but there are some that are at higher risk than others. Young children who have not built up a strong immune system yet are at higher risk along with pregnant women and their fetuses. Elderly people are also at a higher risk because of their lowered immunity. Another group are those that are immunodeficient due to cancer treatment, HIV, or the aids virus. These people should be especially careful and never eat any kind of raw food but all people should take the proper precautions.

BACTERIAL PATHOGENS

Salmonella : Salmonella has traditionally been linked to animal products, including poultry and eggs, although recently it has caused many food-borne illnesses involving fresh fruits and vegetables. Since 2006, Salmonella has been associated with foods such as alfalfa sprouts, tomatoes, jalapeño peppers, black pepper, papayas, cantaloupes, and various nuts. Salmonella is spread via the fecal-oral route, commonly through contact with contaminated water. There are two categories of salmonellosis : non-typhoidal and typhoidal (caused by Salmonella typhi or Salmonella paratyphi). Salmonella strains predominantly cause non-typhoidal salmonellosis in the U.S. Symptoms (commonly diarrhea, nausea, vomiting, abdominal cramps, headache, and fever) typically occur within 72 hours of exposure and last for 4 to 7 days. Most cases are self-limiting, and the mortality rate is less than 1 per cent.

Approximately 8 per cent of patients with non-typhoidal salmonellosis will develop bacteremia and require treatment with antibiotics, including ceftriaxone or azithromycin in children and a fluoroquinolone (commonly levofloxacin) or azithromycin in adults. Patients with non-typhoidal salmonellosis and one or more risk factors for bacteremia should be given antibiotics immediately. These risk factors include age under 3 months or over 65 years, corticosteroid use, inflammatory bowel disease, immunosuppression, and hemodialysis. Antibiotics are recommended in patients with a prosthetic heart valve or abdominal aneurysm to reduce the risk of focal infection. Diarrheal ailments such as salmonellosis are generally treated with rehydration therapy with oral fluids (IV, if necessary), salt intake (commonly through soups and crackers), and an easily digestible diet such as "BRAT" (bananas, rice, applesauce, and toast) until symptoms resolve.

E coli O157 : E coli bacteria are an essential part of the intestinal flora of humans. E coli is usually harmless and, in fact, is protective to the host. However, strains of STEC, such as E coli O157, can cause serious, bloody diarrhea in the host. E coli O157 is predominantly associated with undercooked beef and beef products, although lately it has been more prevalent in fresh produce. In recent years, E coli O157 contamination has occurred in foods such as spinach, romaine lettuce, hazelnuts, and clover sprouts. Waterborne outbreaks have been linked to bodies of fresh water, swimming pools, and ice. After exposure, hemorrhagic colitis often results within 48 hours. Hemorrhagic colitis presents as severe abdominal cramping, nausea, vomiting, and diarrhea that changes from watery to bloody. Life-threatening complications of E coli O157 infections include hemolytic uremic syndrome (HUS) and thrombotic thrombocytopenia purpura. HUS, which occurs when Shiga toxin is absorbed, thereby injuring glomerular cells, results in death in 3 per cent to 5 per cent of cases. Antibiotic therapy is thought to increase the risk of HUS, as is the use of antimotility drugs (*e.g.*, loperamide), which may slow the exit of the enteropathogen from the body; these therapies therefore are not recommended. Treatment should focus on rehydration and on reducing the risk of severe complications.

L monocytogenes : L monocytogenes is another pathogen that can contaminate foods and cause food-borne illness. Unlike many other pathogens, L monocytogenes can live and grow in refrigeration temperatures and salty environments. L monocytogenes contamination has been associated with unpasteurised milk and milk products (*e.g.*, soft cheeses), meats (especially deli meats), hot dogs, raw produce, and seafood. The incidence of L monocytogenes contamination is low, but the fatality rate is high. Pregnant women are at greater risk for listeriosis. Although frequently the mother has mild symptoms, the risks to the fetus are severe. There is a 1-in-3 chance of spontaneous abortion or still-birth, as well as an increased risk of bacteremia or meningitis in the neo-nate.

Listeriosis usually occurs within 3 days after exposure and in less severe cases lasts from days to weeks; symptoms are consistent with a diarrheal ailment. More severe cases may last for several months and can cause complications such as septicemia and meningitis. The treatment of choice for complicated listeriosis is ampicillin (trimethoprim-sulfamethoxazole, in the case of penicillin allergy), with or without gentamicin, for 3 to 6 weeks. As with all food-borne illnesses in which diarrhea is present, rehydration therapy is vital.

Prevention

The key to reducing the incidence of food-borne illness is prevention. Proper food storage, refrigeration, handling, and cooking are vital. Patients should be educated to avoid high-risk items such as unpasteurised milk and milk products, as well as raw or undercooked items like oysters, meat, poultry, and eggs.

The consumption of more meals in the home may also decrease the risk of food-borne illness. According to food-safety experts, two important steps should be implemented in the home to reduce the risk of food-borne illness : washing the hands thoroughly when handling food, and using a thermometer to ensure that adequate food temperatures are reached during cooking. Steps to reduce cross-contamination should also be implemented, including using a separate cutting board and utensils for raw and fresh foods and frequently disinfecting kitchen surfaces. To prevent the spread of bacteria harboured by kitchen towels and sponges, paper towels should be used to clean up cooking spills, and kitchen towels should be laundered frequently. To significantly reduce bacteria, kitchen sponges may be placed in the dishwasher for a heated cycle or microwaved while wet for 1 minute. Some evidence suggests that reusable shopping bags also may harbour bacteria, so separate bags should be used for fresh produce and meat.

Also, washing the bags thoroughly either by hand or in the washing machine may help reduce the spread of bacteria to foods. Using caution to prevent the spread of bacteria is essential to the reduction of food-borne illness.

Table. Recommended Cooking Temperature for Meats, Poultry, and Fish.

Food	Temperature (°F)
Ground beef, pork, veal, lamb	160
Ground turkey, chicken	165
Fresh beef, veal, lamb	145
Fresh chicken, turkey, duck, goose	165
Fresh pork, ham	145
Fin fish	145

A four-step approach to improving food safety, as recommended by the Partnership for Food Safety Education (PFSE). For fresh produce, the PFSE recommends two additional steps. First, be sure that produce is not damaged or bruised and is adequately refrigerated, if precut. Additionally, throw away produce that has been compromised by bruises or improper storage or that may have been in contact with raw meat, poultry, or seafood. When it comes to the safety of fresh produce, the PFSE recommends : "If in doubt, throw it out."

Chapter 19

SOIL BACTERIA AS BIOCONTROL AGENTS AGAINST THE MYCOPARASITIC FUNGI

INTRODUCTION

Use of bacterial flora for management of plant diseases and yield improvement started in the early part of the 20th century. A number of bacterial species have been tested as biocontrol agents. Mechanisms by which bacteria exhibit biological control include antibiosis, competition, parasitism and production of plant growth promoting compounds. Jhune *et. al.* have studied biological control of Mycogone perniciosa, which parasitises Agaricus bisporus. Some of the bacteria isolated from biofertilizers have exhibited a strong antagonism against common competitor moulds. Experiments were conducted to explore and exploit the potential of antagonistic bacteria naturally occurring in casing mixtures for controlling pathogenic fungi of A. bisporus.

Isolation and Identification of Pathogens

Pathogens (Mycogone perniciosa, Verticillium fungicola, Trichoderma harzianum, Fusarium moniliformae and Sepedonium chrysosporium) were isolated from compost/casing/fruit bodies showing symptoms of disease and using peptone dextrose rose bengal agar medium. The cultures of the above pathogens were purified and then their identities were confirmed by comparing their morphological characteristics with the literature. Subsequently purified cultures were maintained on potato-dextrose-agar (PDA) or compost decoction agar (CDA) slants and further multiplied on PDA and CDA media.

Isolation of Bacterial Biocontrol Agents

Employing the serial dilution plate technique, total bacterial colonies were isolated from casing mixtures on soil extract medium and modified King's B medium, respectively. Interaction between bacterial isolates and A. bisporus was studied using a method developed by Weller *et. al.* with slight modification. Five mm disks from a 21-day-old culture of A. bisporus were placed at the centre of petri plates containing compost decoction agar (CDA) medium and then incubated at 22±20C for 20 days. Then a bacterial suspension containing 106 cfu ml-1 was streaked one cm from the growing A. bisporus colony. The plates were incubated for another seven days. Each treatment was replicated three times with a suitable control and observations were made of growth enhancement as well as inhibition.

Screening of the Antagonists

The antagonism between selected bacterial isolates and the pathogenic fungi were tested following the method suggested by Morton and Stroube (1955). Twenty ml of sterilised and melted PDA was poured into 90 mm sterilised petri plates. After solidification, the plates were inoculated with the tested fungus using a five mm diameter disk of young culture. Then a five mm diameter filter paper disk dipped in a bacterial suspension of each individual isolate was placed at a distance of ten mm from the periphery of the inoculated plates. The plates were incubated at 250C for ten days and observations were recorded. Each treatment was replicated three times.

Interaction of Isolates and Pathogens in Mushroom Beds

Five selected isolates were cultured in nutrient broth. Four hundred ml of each isolate containing 106 cfu ml-1 were thoroughly mixed with four kg casing mixture. Inocula of the pathogens, prepared on wheat grains, were mixed into the casing (0.75 per cent of casing weight). Then four cm of casing was applied onto the compost of fully colonised bags. For comparison, there were two controls. In one control, only a pathogen was mixed in casing (Control I); in the other (Control II), neither a pathogen nor an antagonist was used. All the treatments were replicated three times.

MICROBIAL INTERACTIONS AND BIOCONTROL IN THE RHIZOSPHERE

As seeds germinate and roots grow through the soil the loss of organic material provides the driving force for the development of active microbial populations around the root, known as the rhizosphere effect. Although the stimulation in microbial activity is a general phenomenon largely involving saprotrophs, specific groups of symbionts may be selectively enhanced. For example, mutualistic biotrophic symbioses may develop between Rhizobia and legumes, and mycorrhizal fungi may interact with their plant hosts. However, antagonistic symbioses between pathogens and roots can also form resulting in disease. The microbial interactions taking place in the spermosphere and rhizosphere associated with disease development and especially biocontrol of these diseases form the background of this review.

Interest in biological control has increased recently fuelled by public concerns over the use of chemicals in the environment in general, and the need to find alternatives to the use of chemicals for disease control. The key to achieving successful, reproducible biological control is the gradual appreciation that knowledge of the ecological interactions taking place in soil and root environments is required to predict the conditions under which biocontrol can be achieved and, indeed, may be part of the reason why more biocontrol agents are reaching the market-place. This type of work requires a study not only of any potential biocontrol agent *per se* but also its interactions with the crop, the natural resident microbiota and the environment as well.

In this regard, it is well known that some soils are naturally suppressive to some soil-borne plant pathogens such as Fusarium oxysporum Schlect. : Fr. Emend. Snyder and Hansen, Gaeumannomyces graminis (Sacc.) v. Arx and Oliver, Pythium and Phytophthora species and this suppression relates to both physicochemical and microbiological features of the soil. Importantly, a soil that is suppressive to one pathogen is not necessarily suppressive to another, and so specificity in the soil–plant–microbe interactions for disease suppression exists.

Modern methods for analysing microbial community structures may prove particularly valuable to help define the key organisms or groups of organisms responsible for such natural suppression as well as for monitoring the spread and impact of introduction of specific biocontrol agents or other management practices on natural microbial populations. Significantly, disease suppression can also be achieved by manipulation of the physico-chemical and microbiological environment through management practices such as the use of soil amendments, crop rotations, use of fumigants or soil solarisation. However, at present, greatest interest resides with the development and application of specific biocontrol agents for the control of diseases on seeds and roots and the interaction of these with pathogens and hosts, and will form the focus of this paper. There have been numerous reviews in recent years on this topic and so only selected recent examples, illustrating key features of biocontrol on seeds and roots, particularly the different modes of action, will be discussed whenever possible.

Modes of action include : inhibition of the pathogen by anti-microbial compounds (antibiosis); competition for iron through production of siderophores; competition for colonisation sites and nutrients supplied by seeds and roots; induction of plant resistance mechanisms; inactivation of pathogen germination factors present in seed or root exudates; degradation of pathogenicity factors of the pathogen such as toxins; parasitism that may involve production of extracellular cell wall-degrading enzymes, for example, chitinase and ß-1,3 glucanase that can lyse pathogen cell walls. None of the mechanisms are necessarily mutually exclusive and frequently several modes of action are exhibited by a single biocontrol agent. Indeed, for some biocontrol agents, different mechanisms or combinations of mechanisms may be involved in the suppression of different plant diseases.

Bacteria–Bacterial Pathogen Interactions

In the last few years there have been relatively few studies of bacteria applied to seeds and roots for the purpose of controlling bacterial diseases. One example, is the application of non-pathogenic strains of Streptomyces to control scab of potato (Solanum tuberosum L.) caused by Streptomyces scabies (Thaxter) Waksman and Henrici. Here biocontrol may operate through antibiosis or competition for space or nutrients in the rhizosphere.

In contrast, Pseudomonas fluorescens (Trevisan) Migula F113 was shown to control the soft rot potato pathogen Erwinia carotovora subsp. atroseptica (van Hall) Dye by production of the antibiotic 2,4-diacetylphloroglucinol (DAPG) and, through use of co-inoculation experiments with mutants lacking DAPG production, that competition was not a feature of biocontrol in this system. Some evidence was also obtained that siderophore production by P. fluorescens F113 may play a role in biocontrol of potato soft rot under iron-limiting conditions, but DAPG appears to be the major biocontrol determinant. Pseudomonas species may also control crown gall disease in many dicotyledonous plants caused by Agrobacterium tumefaciens (Smith and Townsend) Conn.

However, the classic, and still commercially successful, bacterial-based biocontrol system is the use of non-pathogenic Agrobacterium strains to control Agrobacterium tumefaciens. Long-term molecular and ecological studies of this control system have identified how the biocontrol works and have also allowed potential problems associated with its use in the field to be overcome. The most widely used non-pathogenic Agrobacterium strain K84 produces a highly specific antibiotic agrocin 84, which is encoded by plasmid pAgK84. Inundative inoculation of Agrobacterium strain K84 to roots by dipping in cell suspensions prior to exposure to the pathogen effectively controls those strains of pathogen susceptible to agrocin 84. However, because there is a risk that plasmid pAgK84 could be transferred to pathogenic strains and reduce effectiveness of control, a transfer deletion mutant of K84, K1026 has been constructed. Strain K1026 is as efficient as K84 in biocontrol of both strains susceptible to agrocin 84 and those resistant to agrocin 84 and so, clearly, production of agrocin 84 is not the only mechanism of biocontrol. Production of other antibiotics such as agrocin 434 or ALS 84 may play a part, but the ability to survive and compete on roots may also be important. Studies where pathogenic cells were co-inoculated with K84 or K1026 resulted in survival of the pathogen on roots up to 8 months later, although no symptoms were present, providing evidence that the non-pathogenic strains prevented disease expression rather than killing pathogen cells directly.

Bacteria–Fungal Pathogen Interactions

The volume of literature in this area continues to increase at a rapid rate, stimulated by the increasing ease with which molecular techniques can be applied to answer questions concerning distribution, and occurrence and relative importance of specific modes of action. Some examples of the different types of bacteria-fungal pathogen interactions examined in the spermosphere and rhizosphere in just the last three years are given in Table? Although a range of different bacterial genera and species have been studied, the overwhelming number of papers have involved the use of Pseudomonas species. Clearly, Pseudomonas species must have activity but it begs the question as to the features that make this genus so effective and the choice of so many workers. Pseudomonads are characteristically fast growing, easy to culture and manipulate genetically in the laboratory, and are able to utilise a range of easily metabolisable organic compounds, making them amenable to experimentation. But, in addition, they are common rhizosphere organisms and must be adapted to life in the rhizosphere to a large extent. Having appropriate ecological rhizosphere competence may be a key feature for reproducible biological control activity in the spermosphere and rhizosphere. This criterion is already widely appreciated for many fungal biocontrol agents. A few specific examples of the modes of action involved with bacterial biocontrol of fungal pathogens in the rhizosphere and spermosphere are given below.

Antibiosis

There are numerous reports of the production of antifungal metabolites (excluding metal chelators and enzymes) produced by bacteria in vitrothat may also have activity *in vivo*. These include ammonia, butyrolactones, 2,4-diacetylphloroglucinol (Ph1), HCN, kanosamine, Oligomycin A, Oomycin A, phenazine-1-carboxylic acid (PCA), pyoluterin (Plt), pyrrolnitrin (Pln), viscosinamide, xanthobaccin, and zwittermycin A as well as several other uncharacterised moieties. To demonstrate a role for antibiotics in biocontrol, mutants lacking production of antibiotics or over-producing mutants have been used. Alternatively, the use of reporter genes or probes to demonstrate production of antibiotics in the rhizosphere is becoming more commonplace. Indeed, isolation and characterisation of genes or gene clusters responsible for antibiotic production has now been achieved. Significantly, both Phl and PCA have been isolated from the rhizosphere of wheat following introduction of biocontrol strains of Pseudomonas, finally confirming that such antibiotics are produced *in vivo*. Further, Ph1 production in the rhizosphere of wheat was strongly related to the density of the bacterial population present and the ability to colonise roots. PCA from Pseudomonas aureofaciens Kluyver Tx-1 has even been used as a direct field treatment for the control of dollar spot (Sclerotinia homeocarpa F. T. Bennett) on creeping bentgrass (Agrostis palustrisHudson).

Antibiotic production by bacteria, particularly pseudomonads, seems to be closely regulated by a two-component system involving an environmental sensor (presumably a membrane protein) and a cytoplasmic response factor. Mutation in either gene has similar multiple effects on antibiotic production. For example, P. fluorescens Pf-5 with a mutation in the apdA sensor gene lost the ability to produce HCN, Plt and Pln and P. fluorescens CHA0 with a defect in the gacA response gene lost the ability to produce Phl, Plt and HCN as well as protease and phospholipase C. However, the environmental signals that control the two-component system are unknown. Interestingly, the gacA gene is required for biocontrol activity in P. fluorescens CHA0 in the rhizosphere of dicotyledous plants, but not in those of the Gramineae, although the mechanisms are unclear.

Other two-component signalling phenomena may also be involved in PCA production by pseudomonads on roots. Pseudomonas aureofaciens 30-84 is a biocontrol agent of take-all disease of wheat (Triticium aestivum L.) caused by Gaeumannomyces graminis var. tritici Walker and operates through inhibiting growth of the pathogen by production of PCA. In this system, pathogen growth on the root increases root exudation and this results in an increase in the population of P. aureofaciens 30-84 and other bacteria in the infection zone. Consequently, there is an increase in the level of the signal molecule N-acyl-L-homoserine lactone (HSL) produced at low levels constitutively by the phzI gene, in the rhizosphere which is sufficient to switch on the PCA synthesis pathway in P. aureofaciens 30-84 controlled by the phzR gene. The resulting PCA production inhibits further growth of the pathogen. This explains why P. aureofaciens 30-84 does not reduce the number of infection sites on the roots, but inhibits secondary growth of the pathogen. Significantly, HSL from other members of the rhizosphere microbial community can contribute to PCA production in P. aureofaciens 30-84 raising the question of the significance of interpopulation signalling on biocontrol and perhaps the enhanced performance of certain strains of bacteria when introduced with mixtures of other bacterial biocontrol strains. In addition, antibiotic production in Pseudomonas spp. may be further controlled by the activity of housekeeping sigma factors encoded by rpoS or rpoD genes, illustrating the complexity of these regulatory systems.

Interestingly, signalling between pathogenic fungi and potential biocontrol bacteria has also been detected. In one case, trehalose derived from Pythium debaryanum Hesse up-regulated genes in its biocontrol strain Pseudomonas fluorescens ATCC 17400 and yet in another example Pythium ultimum Trow caused a down-regulation of five gene clusters of P. fluorescens F113 which provides biocontrol of this pathogen in the rhizosphere of sugar beet. These findings may be of considerable significance for bacterial–fungal interactions in general and has major implications for the control of gene expression in complex microbial communities.

Competition for Iron

Although competition between bacteria and fungal plant pathogens for space or nutrients has been known to exist as a biocontrol mechanism for many years the greatest interest recently has involved competition for iron. Under iron-limiting conditions, bacteria produce a range of iron chelating compounds or siderophores which have a very high affinity for ferric iron. These bacterial iron chelators are thought to sequester the limited supply of iron available in the rhizosphere making it unavailable to pathogenic fungi, thereby restricting their growth.

Recent studies have clearly shown that the iron nutrition of the plant influences the rhizosphere microbial community structure. Iron competition in pseudomonads has been intensively studied and the role of the pyoverdine siderophore produced by many Pseudomonas species has been clearly demonstrated in the control of Pythium and Fusarium species, either by comparing the effects of purified pyoverdine with synthetic iron chelators or through the use of pyoverdine minus mutants. Pseudomonads also produce two other siderophores, pyochelin and its precursor salicylic acid, and pyochelin is thought to contribute to the protection of tomato plants from Pythium by Pseudomonas aeruginosa (Schroeter) Migula 7NSK2. However, siderophores are not always implicated in disease control by pseudomonads. The dynamics of iron competition in the rhizosphere are often complex. For example, some siderophores can only be used by the bacteria that produce them, whereas others can be used by many different bacteria. Different environmental factors can also influence the quantity of siderophores produced. There is also the further complication that pyoverdine and salicylate may act as elicitors for inducing systemic resistance against pathogens in some plants.

Parasitism and Production of Extra-cellular Enzymes

The ability of bacteria, especially actinomycetes, to parasitise and degrade spores of fungal plant pathogens is well established. Assuming that nutrients pass from the plant pathogen to bacteria, and that fungal growth is inhibited, the spectrum of parasitism could range from simple attachment of cells to hyphae, as with the Enterobacter cloacae (Jordan) Hormaeche and Edwards–Pythium ultimum interaction to complete lysis and degradation of hyphae as found with the Arthrobacter–Pythium debaryanum interaction. If fungal cells are lysed and cell walls are degraded then it is generally assumed that cell wall-degrading enzymes produced by the bacteria are responsible, even though antibiotics may be produced at the same time. Considerable effort has gone into identifying cell wall-degrading enzymes produced by biocontrol strains of bacteria even though relatively little direct evidence for their presence and activity in the rhizosphere has been obtained. For example, biocontrol of Phytophthora cinnamomi Rands root rot of Banksia grandis Willd. was obtained using a cellulase-producing isolate of Micromonospora carbonacea Luedemann and Brodsky and control of Phytophthora fragariaevar. *rubi* Hickman causing raspberry root rot was suppressed by the application of actinomycete isolates that were selected for the production of β-1,3-, β-1,4- and β-1,6-glucanases.

Chitinolytic enzymes produced by both Bacillus cereus Frankland and Pantoea (Enterobacter) agglomerans (Beijerinck) Gavini *et. al.* also appear to be involved in biocontrol of Rhizoctonia solani Kühn. Tn5 mutants of E. agglomerans (Beijerinck) Gavini *et. al.* deficient in chitinolytic activity were unable to protect cotton (Gossypium barbardense L.) and expression of the chiA gene for endochitinase in Escherichia coli (Migula) Castellani and Chalmers allowed the transformed strain to inhibit R. solani on cotton seedlings. Similar techniques involving Tn5 insertion mutants and subsequent complementation demonstrated that biocontrol of Pythium ultimum in the rhizosphere of sugar beet by Stenotrophomonas maltophila (Hugh) Palleroni and Bradbury W81 was due to the production of extracellular protease.

Induced Resistance

Perhaps the greatest growth area in biocontrol in the last few years has been concerned with induced resistance defined as the process of active resistance dependent on the host plant's physical or chemical barriers, activated by biotic or abiotic agents (inducing agents). This has come about through the synergistic interaction of microbiologists, plant pathologists and plant scientists armed with an appropriate battery of molecular tools. The effect had previously often been overlooked through inadequate techniques or controls as well as the biocontrol agent exhibiting other modes of action at the same time. Most work has focused on the systemic resistance induced by non-pathogenic rhizosphere-colonising Bacillus and Pseudomonas species in systems where the inducing bacteria and the challenging pathogen remained spatially separate for the duration of the experiment, and no direct interaction between the bacteria and pathogen was possible. Such split root or spatial root inoculation experiments were used to demonstrate the phenomenon in radish (Raphanus sativus L.) and Arabidopsis against Fusarium oxysporum and in cucumber (Cucumis sativus L.) against Pythium aphanidermatum (Edson) Fitzp. Various combinations of timing and position have indicated that induced resistance also occurs in carnation (Dianthus caryophyllus L.), tobacco (Nicotiana tabacum L.) and tomato (Lycopersicon esculentum Mill.). Bacteria differ in ability to induce resistance, with some being active on some plant species and not others; variation in inducibility also exists within plant species. The full range of inducing moieties produced by bacteria is probably not yet known, but lipopolysaccharides and siderophores are clearly indicated.

The definition of induced resistance suggested by Kloepper *et. al.* covered both biotic and abiotic inducers. Although the phenotypic effects of root inoculation with bacteria may be similar to treatment with abiotic agents or micro-organisms that cause localised damage, the biochemical and mechanistic changes appear to be subtly different. This has resulted in the term induced systemic resistance (ISR) for bacterially-induced resistance and systemic acquired resistance for the other forms. The major differences are that pathogenesis-related (PR) proteins such as chitinases, b-1,3-glucanases, proteinase inhibitors and one or two other rarer types, are not universally associated with bacterially induced resistance and salicylic acid (a known inducer of SAR) is not always involved in expression of ISR, but this is dependent on bacterial strain and host plant involved. Ethylene responsiveness may also be required at the site of inoculation of the inducing bacteria for ISR to occur.

Changes that have been observed in plant roots exhibiting ISR include :

- Strengthening of epidermal and cortical cell walls and deposition of newly formed barriers beyond infection sites including callose, lignin and phenolics;
- Increased levels of enzymes such as chitinase, peroxidase, polyphenol oxidase, and phenylalanine ammonia lyase;
- Enhanced phytoalexin production;
- Enhanced expression of stress-related genes.

However, not all of these biochemical changes are found in all bacterial–plant combinations. Similarly, the ability of bacteria to colonise the internal tissue of the roots has been considered to be an important feature in many of the bacterial–root interactions involving ISR, but is not a constant feature of them all.

Plant Growth-Promoting Rhizobacteria (PGPR)

The concept of PGPR is now well established and so some consideration of the relationship of PGPRs to biocontrol is worthwhile. PGPR increase plant growth indirectly either by the suppression of well-known diseases caused by major pathogens or by reducing the deleterious effects of minor pathogens (micro-organisms which reduce plant growth but without obvious symptoms). Most of the bacteria discussed so far in this review fall into this category of PGPR.

Alternatively, PGPR may increase plant growth in other ways, for example, by associative N2 fixation, solubilising nutrients such as P, promoting mycorrhizal function, regulating ethylene production in roots, releasing phytohormones, and decreasing heavy metal toxicity.

It has been suggested that the two groups should be reclassified into biocontrol plant growth-promoting bacteria (biocontrol PGPB) and PGPB. To date this proposal does not seem to have been widely accepted, but it does highlight the need to consider the full ecological interactions taking place following application of bacteria to seeds and roots that lead to plant growth promotion. It is also important to remember that deleterious rhizobacteria that inhibit plant growth are also known which can influence such interactions.

Irrespective of mode of action, a key feature of all PGPR is that they all colonise roots to some extent. In some cases this may involve specific attachment through, for example, pili, as with the attachment of Pseudomonas fluorescens 2-79 to the surface of wheat roots. However, such specific attachment does not seem to be an absolute requirement for colonisation. Colonisation may involve simply root surface development but, endophytic colonisation of the root is also known, and the degree of endophytic colonisation depends on bacterial strain and plant type. Endophytic growth in roots has been recorded with the PGPR Bacillus polymyxa (Prazmowski) Macé Pw-ZR and Pseudomonas fluorescens Sm3-RN on spruce (Picea glauca×P. engelmannii), with the biocontrol strains of Bacillus sp. L324-92R12 and P. fluorescens 2-79RN10 on wheat and with several that induce resistance such as Bacillus pumilus Meyer and Gottheil SE34 and P. fluorescens 63-28 on pea (Pisum sativum L.), P. fluorescens CHA0 on tobacco and P. fluorescensWCS417r on tomato. Large scale differences in spread within the plant may occur.

Some, such as B. polymyxa Pw-2B, P. fluorescens SM3-RN, Bacillus sp. L324-92R12, and P. fluorescens 2-79RN10spread from roots to aerial plant parts whereas others may not. Small scale differences are also known. For example, both B. pumilus SE34 and P. fluorescens 63-28 grow on the root surface and intercellarly in pea roots whereas surface growth, inter- and intra-cellular growth occurred with P. fluorescens WCS417r in tomato and P. fluorescens CHA0 in tobacco. These endophytic bacteria may be in a particularly advantageous ecological position in that they may be able to grow and compete on the root surface, but also may be capable of developing within the root, relatively protected from the competitive and high-stress environment of the soil.

Indeed, many seeds, roots and tubers are normally colonised by endophytic bacteria. Any plant resistance encountered must be minimal, although, in many cases, sufficient to allow ISR to develop. The localised signalling between plant and bacteria within the root environment deserves further study. Certainly, use of mutants and promoter probe techniques are beginning to identify genes in bacteria that are important in colonisation and these are often related to nutrient uptake. Such nutrient uptake genes may also play a role in biocontrol by aiding the uptake and metabolism of nutrients that stimulate germination of pathogen propagules.

The ability to colonise seeds is also an important feature for many bacterial biocontrol agents. Pseudomonas chlororaphis (Guignard and Sauvngeau) Bergey *et. al.* MA342 is applied to cereal seeds to control many seed and soil-borne pathogens and has been found to colonise specific areas of the seed coat. After inoculation, the bacteria were found under the seed glume (or husk), but after planting they were found to colonise the glume cells epiphytically. Bacterial aggregates were also found in the grooves formed by the base of the coleoptile and the scutellum, and near the embryo but never within it.

In this case, the biocontrol bacteria co-located with the seed-borne pathogen Drechslera teres (Sacc.) Shoemaker providing biocontrol through the production of fungitoxic compounds. The spermosphere competence of this bacterium allowed biocontrol to take place. Microcolony or microaggregate production by bacteria has also been found on the grooves or cracks on the outer seed coat of sugar beet and cotton (Gossypium hirsutum L.) perhaps reflecting areas of increased nutrient availability or environmental protection.

Fungal-Protozoan Interactions

The soil-borne protozoan Plasmodiophora brassicae Woronin is an ecologically obligate biotroph of brassicas causing clubroot disease which is characterised by proliferation of galls on infected roots. From a large-scale screening exercise, two isolates of the root-colonising fungus Heteroconium chaetospira (Grove) Ellis were found to suppress clubroot on chinese cabbage (Brassica campestris L.) in non-sterile soil. Hyphal growth occurred in the inner parts of the cortical tissues and into the root tips without causing any external symptoms on the plant and there was no sign of infection by P. brassicae. Further studies demonstrated that H. chaetospira infected epidermal cells from appressoria via infection pegs and, subsequently, intracellular hyphal growth occurred. However, the actual mechanism of the disease control observed in the field was unclear. Heteroconium chaetospira appears to form a mutualistic symbiosis with B. campestris in terms of disease control which is of interest as the Brassicacae family is largely non-mycorrhizal.

In addition, H. chaetospira was found to colonise the roots of plants from eight families and may have a wide host range. Its ability to control diseases in these other plant families and the mechanisms involved deserves further study.

Fungal–Bacterial Pathogen Interactions

In the last few years there have been no clear examples of fungi used to control bacterial plant pathogens in the rhizosphere or spermosphere. The reasons for this are unclear but could perhaps indicate an area that deserves further research in the future.

Fungal–Fungal Pathogen Interactions

Interactions between biocontrol fungi and fungal plant pathogens continue to be the focus of a large number of researchers, on a par with work on bacterial–fungal plant pathogen interactions described earlier. However, there is an extra dimension in the quality of the interactions between fungi as biocontrol fungi have much greater potential than bacteria to grow and spread through soil and in the rhizosphere through possession of hyphal growth. Some recent examples of fungal–fungal interaction concerning biocontrol in the rhizosphere and spermosphere.

There are a variety of fungal species and isolates that have been examined as biocontrol agents but Trichoderma species clearly dominate, perhaps reflecting their ease of growth and wide host range. There has been an upsurge in interest in non-pathogenic Pythium species, particularly P. oligandrum Drechsler where additional modes of action have been determined recently, and a continued interest in well-established saprotrophic antagonists such as non-pathogenic Fusarium species, non-pathogenic binucleate Rhizoctoniaisolates and Phialophora species, as well as mutualistic symbionts including mycorrhizal fungi such as Glomus intraradices Schenk and Smith. At least one novel biocontrol agent, Cladorrhinum foecundissimum Saccardo and Mardial, has been described. Numerous others are listed elsewhere. The most common pathogen targets are Pythium species, Fusarium species and Rhizoctonia solani reflecting their world-wide importance and perhaps their relative ease of control under protected cropping systems, although numerous other pathogens have been examined. Significantly, involve studies in non-sterile soil or field conditions, with most carried out in soil-less conditions reflecting the need to keep the complexity of the system to a minimum in order to achieve reproducible control. Some specific examples of the modes of action found to occur in the rhizosphere and spermosphere during interactions between fungi and fungal plant pathogens.

Competition

There have been relatively few studies on competition for nutrients, space or infection sites between fungi in the rhizosphere and spermosphere recently. Competition for carbon, nitrogen and iron has been shown to be a mechanism associated with biocontrol or suppression of Fusarium wilt in several systems by non-pathogenic Fusarium and Trichoderma species and competition for thiamine as a significant process in the control of Gaeumannomyces graminis var. tritici by a sterile red fungus in the rhizosphere of wheat. Many studies have shown a relationship between increased colonisation of the rhizosphere by a non-pathogen, associated subsequently, with disease suppression.

This is well established for non-pathogenic strains of Fusarium oxysporum-controlling pathogenic F. oxysporum on a variety of crop plants, hypovirulent or non-pathogenic binucleate strains of Rhizoctonia species to control pathogenic isolates of R. solani and several fungi including Phialophora species, Gaeumannomyces graminis var. graminis and Idriella bolleyi (Sprague) von Arx as well as several non-sporulating fungi, to control G. graminis var. tritici. As just one example, I. bolleyi exploits the naturally senescing cortical cells of cereal roots during the early stages of the crop and outcompetes G. graminis var. tritici for infection sites and nutrients. Rapid production of spores, which are then carried down the root by water, continue the root colonisation process and this is suggested to be a key feature in the establishment of the biocontrol agent on the root.

Mycorrhizal fungi are also strong candidates for providing biocontrol through competition for space by virtue of their ecologically obligate association with roots. Ectomycorrhizal fungi because of their physical sheathing morphology may well occupy normal pathogen infection sites. Strangely, little work has been carried out to demonstrate this mechanism since it was first suggested with most biocontrol interest focused on antibiotic production and induced resistance. Similarly, arbuscular mycorrhizas also have potential to occupy space and infection sites on roots, but evidence suggests that biocontrol provided by arbuscular mycorrhizas relates more to induced resistance, improved plant growth and changes in root morphology rather than competition *per se.*

Antibiosis

Although production of antibiotics by fungi involved in biocontrol is a well-documented phenomenon, there is little recent work clearly demonstrating production of antibiotics by fungi in the rhizosphere and spermosphere. Unlike the situation with biocontrol bacteria, there appear to be no detailed studies in biocontrol fungi of genes coding for antibiotic synthesis.

Mutants with raised or decreased production of antibiotics are either natural spontaneous ones or generated by UV or chemical mutagenesis, with inherent problems of pleiotropic gene effects, rather than targeted gene disruption. Consequently, clear identification and understanding of the role of antibiotics in disease control lags far behind that in bacteria and needs to be addressed.

Antibiotic production by fungi exhibiting biocontrol activity has most commonly been reported for isolates of Trichoderma/Gliocladium and Talaromyces flavus (Klöcker) Stolk and Samson although in the last few years antibiotics have been at least partially characterised in Chaetomium globosum (Kunze). Minimedusa polyspora Weresub and Le Clair and Verticillium biguttatum Gams. Of particular interest are those studies where antibiotic production has a definite link to biocontrol. For example, Trichoderma (Gliocladium)virens (J. Miller, Giddens and Foster) von Arx comprises P and Q group strains, based on their antibiotic profiles.

Strains of P group produce the antibiotic gliovirin which is active against Pythium ultimum but not against Rhizoctonia solani AG-4. Strains of the Q group produce the antibiotic gliotoxin which is very active against R. solani but less so against P. ultimum. In seedling bioassay tests, strains of the P group are more effective biocontrol agents of damping-off on cotton caused by Pythium, while those from the Q group are more effective as biocontrol agents of damping-off incited by R. solani. Thus there is strong circumstantial evidence for a role for antibiotics in biocontrol in this experimental system.

This has been confirmed in a zinnia–Pythium system where T. virens G-20 incorporated into soil and potting mix resulted in disease suppression clearly associated with maximum accumulation of gliotoxin in the medium. Gliotoxin minus mutants displayed only 54 per cent of the Pythium disease suppressive activity in zinnia compared with the wild-type. Gliotoxin production by Trichoderma is also thought to be responsible for cytoplasmic leakage from R. solani observed directly on membranes in potting mix.

Production of hydrogen peroxide in the rhizosphere, catalysed by glucose oxidase from Talaromyces flavus is thought to be responsible for the biocontrol of Verticillium wilt caused by Verticillium dahliae Kleb. on eggplant (Solanum tuberosum L.). Purified glucose oxidase significantly reduced the growth rate of V. dahliae in the presence, but not the absence, of eggplant roots, suggesting that a supply of glucose from the roots was of major importance. Further, a single-spored variant, Tf-l-np, which produced 2 per cent of the level of glucose oxidase activity of the wild-type did not control Verticillium wilt on eggplant in non-sterile field soil in a glasshouse experiment, whilst the wild-type provided significant control. Glucose oxidase also suppressed growth of V. dahliae *in vitro* and killed microsclerotia of V. dahliae in vitro and in soil.

Induced Resistance

As with bacteria described earlier, the ability of fungi to induce resistance in plants and provide biocontrol has gradually been receiving more attention in the last few years. A considerable number of fungi previously described to provide biocontrol by mechanisms such as competition, antibiosis, mycoparasitism or direct growth promotion are now thought to provide control, at least in part, by this mechanism. These include saprotrophs such as non-pathogenic Fusarium isolates, Trichoderma species, Pythium oligandrum, non-pathogenic binucleate Rhizoctonia isolates, and Penicillium oxalicum Currie and Thom as well as mutualistic biotrophs such as mycorrhizal fungi.

However, not all these studies used the strict criterion of spatial separation between application of the biocontrol fungus and the challenging pathogen to define induced resistance. Indeed, some simply measured changes in enzymes, PR-proteins or cell wall characteristics found to be induced in plants through SAR without involvement of a pathogen at all. Certainly with some mycorrhizal fungi it has been questioned whether the biochemical responses similar to induced resistance found following infection are of sufficient magnitude or quality, or too transient, to provide disease control. Indeed, during some mycorrhizal syntheses there is little or no induced resistance response detected. However, spatial or temporal separation experiments have indicated that increased levels of chitinases, b-1,3 glucanases, b-1,4 glucosidase, PR-1 protein, and peroxidase as well as cell wall appositions and phenolics may be associated with induced resistance due to fungi. Nevertheless, more work is needed to identify the biochemical changes taking place in a larger number of fungal–plant combinations as not all these biochemical markers were found to be important in each system examined. Further, there appear to be differences in the quality of the induced resistance found between bacteria and on fungi on the same plant. For example, the suppression of Fusarium wilt Snyder and Hansen on tomato by Pseudomonas fluorescens WCS417r did not involve production of PR-1 protein and chitinases whereas that induced by F. oxysporum Fo47 did. Again more work in this area is required to determine the extent of differences in induced resistance produced by bacteria and fungi.

The elicitors responsible for inducing resistance are not known in detail.Trichoderma species produce a 22?kDa xylanase that, when injected in plant tissues, will induce plant defence responses including K+, H+ and Ca2+ channelling, PR protein synthesis, ethylene biosynthesis, and glycosylation and fatty acylation of phytosterols. However, whether such a system is active in roots exposed toTrichoderma is not known. Pectic oligogalacturonides released after hydrolysis by a non-pathogenic binucleate Rhizoctonia isolate may act as elicitors of defence responses in bean.

Dose–response experiments involving non-pathogenic Fusarium species to control F. oxysporum on tomato have indicated that induced resistance is not an all or nothing response. By varying the level of inoculum of the inducing strain and the pathogenic isolate in soil, it was shown that some non-pathogenic isolates such as Fusarium CS-20 controlled Fusarium wilt effectively with antagonist levels of only 100 chlamydospores g-1 of soil (cgs) with pathogen densities of up to 105?cgs. In contrast, isolate Fo47 was effective only at antagonist densities of 104–105 cgs, regardless of pathogen density. Subsequent mathematical modelling provided evidence that CS-20 control was largely through induced resistance whereas Fo47 was active primarily through competition for nutrients. Similar dose–response effects were found with non-pathogenic isolate of F. oxysporum f. sp.ciceris (Padw.) Matuo and Sato and non-pathogenic isolates of F. oxysporumto control wilt of chickpea (Cicer arietinum L.) caused by pathogenic F. oxysporum f. sp. ciceris. However, in addition, the plant genotype also seemed to influence the degree of resistance induced.

Mycoparasitism

There is a huge literature on the ability of fungi to parasitise spores, sclerotia, hyphae, and other fungal structures and many of these observations are linked with biocontrol. However, most of the microscopical observations concerning mycoparasitism have come from *in vitro* studies or sterile systems and examples clearly demonstrating mycoparasitism in the rhizosphere or spermosphere are rare. However, indirect population dynamic studies showed that mycelium of Rhizoctonia solani in the rhizosphere of potato was a pre-requisite for development of the mycoparasite Verticillium biguttatum and rhizosphere competence was strongly related to biocontrol in mycoparasite isolates of Trichoderma species.

The process involved in mycoparasitism may consist of sensing the host, followed by directed growth, contact, recognition, attachment, penetration, and exit. Although not all these features occur in every fungal–fungal interaction, the key factor is nutrient transfer from host to mycoparasite. Directed growth of hyphae of Trichoderma to hyphae of Rhizoctonia solani prior to penetration has often been observed and the presence of host sclerotia have been shown to stimulate germination of conidia of Coniothyrium minitans Campbell and Sporidesmium sclerotivorum Uecker, Adams and Ayers.

However, the factors involved in controlling directed growth in these systems have not been fully characterised. Similarly, the factors controlling recognition and binding between fungal host and parasite are not yet clear. This process may involve hydrophobic interactions or interactions between complementary molecules present on the surface of both the host and the mycoparasite such as between lectins and carbohydrates. With Trichoderma, there is good evidence of lectin production by both parasite and host Corticum (Sclerotium) rolfsii Curzi and involvement of lectins in the differentiation of mycoparasitism-related structures. Recently, both hydrophobic characteristics and surface carbohydrate moieties have been investigated in the mycoparasite C. minitans as a prerequisite to examining the interaction with its host Sclerotinia sclerotiorum (Lib.) de Bary. Little is known of the signalling pathways following recognition of the host. However, preliminary evidence in Trichoderma harzianum indicates that the signal is transduced by heterotrimeric G proteins and mediated by cAMP.

As penetration or cell wall degradation are frequently observed during mycoparasitism, great emphasis has been placed on characterising and cloning the extracellular enzymes such as b-1,3 glucanases, chitinases, cellulases, and proteases produced by fungal biocontrol strains. By manipulating their activity through construction of 'overproducing' mutants, enzyme-negative mutants or even transgenic plants expressing the enzyme, a role for their production in biocontrol has been implied. Several fungi have been examined in this way including Talaromyces flavus, but this type of work has essentially focused on Trichoderma species. For example, a series of transformants of T. longibrachiatum Rifai were constructed with extra copies of egl1 gene encoding the production of β-1,4 endoglucanase. When applied to cucumber seeds sown in Pythium ultimum-infested soil, the transformants with inducible or constitutive expression of egl1 were generally more suppressive than the wild type strain. In this case, it was suggested that P. ultimum was controlled through the action of β-1,4 glucanase degrading the cellulose of the cell wall of the pathogen.

Similarly, transformants of T. harzianum, overproducing proteinase encoded by prb1, provided up to a 5-fold increase in control of damping-off in cotton caused by Rhizoctonia solani. Interestingly, the best protection was provided by a strain which produced only an intermediate level of proteinase activity and it was suggested that very high levels of proteinase production might cause degradation of other enzymes which are important in the mycoparasitic process. In this regard, chitinases have received the greatest attention in mycoparasitism.

Numerous studies have been made of β-N-acetylhexosaminidase, endochitinase and chitin 1,4-β-chitobiosidase (exochitinases or chitobiosidases) (which releases diacetylchitobiose in a stepwise fashion such that no monosaccharides or oligosaccharides are formed). For example, transformants of T. harzianum CECT 2413 that over-expressed on 33 kDa endochitinase (chit33) were more effective in inhibiting the growth of Rhizoctonia solani *in vitro* compared with the wild type.

The combination of chitinases as well as other cell wall-degrading enzymes differ between species and strains and chitinases are differently expressed during mycoparasitism. For example, an N-acetylhexosaminidase was the first to be induced in T. harzianum T-Y, but as early as 12 h after contact with its host Sclerotium rolfsii, its activity diminished, while that of another N-acetylhexosaminidase increased. In contrast, when Rhizoctonia solani was the host, CHIT 102 was stimulated along with three endochitinases within 12 h following contact but, as the interaction proceeded, CHIT 102 activity decreased and that of the endochitinases increased. In T. (atroviride) harzianum P1, expression of endochitinase ech42 occurred before contact with the hostR. solani when a N-acetylhexosaminidase nag1 was not induced until after contact. Degradation products of the cell wall were considered to act as inducers of these enzymes in this host–pathogen system. Interestingly, when examining T. harzianum P1 interaction with Botrytis cinerea Pers., ech42 gene transcription was found to be triggered by physiological stress reflecting carbon source depletion rather than the presence of chitin as the inducer. Significantly, ech42 minus mutants did not reduce biocontrol activity against Pythium ultimumor S. rolfsii and appeared to enhance activity against R. solani, suggesting that expression of this gene in T. harzianum P1 was not directly important for the biocontrol of these fungi. In contrast, activity against B. cinerea was impaired, clearly indicating that the mechanism involved in antagonistic interactions with T. harzianum P1 differs with host, and is unlikely to depend solely on the production of a single cell wall-degrading enzyme. Indeed, evidence from studies involving combinations of purified cell wall-degrading enzymes and antibiotics from Trichoderma species support this idea.

The final evidence for a role for cell wall-degrading enzymes in biocontrol involves the expression of fungal genes in transgenic plants. For example, an endochitinase from Trichoderma harzianum has been transformed into tobacco and potato and the transgenic plants showed a high level of resistance to a broad spectrum of diseases. Similarly, transgenic apple trees expressing an endochitinase from T. harzianumalso exhibited increased resistance to apple scab caused by Venturia inaequalis (Cooke) Winter, although plant growth was reduced. The potential consequently exists to combine different enzymes in transgenic plants to obtain synergistic biocontrol and these experiments are underway.

One feature often overlooked with mycoparasitism is that it may not always be confined to control of plant pathogens. A mycoparasite may also have the potential to attack beneficial fungi such as those forming mycorrhiza. For example, T. harzianum T-203 was shown to attack mycelium of the arbuscular mycorrhizal fungus Glomus intraradices in an axenic system. However, in a soil-based system, G. intraradices was unaffected by the presence of T. harzianum T3a. IndeedT. harzianum appeared to be suppressed through nutrient competition. In contrast, in a different *in vitro* system, soluble material obtained from G. intraradices mycelium stimulated conidial germination of T. harzianum, but not that of the pathogen F. oxysporumf. sp. chrysanthemi G. M. and J. K. Armstrong and Littrell. These results may reflect the different isolates of fungi and plants used, and the experimental systems applied, but they clearly demonstrate the complexity of the interactions that can occur in the rhizosphere.

Plant Growth Promotion and Rhizosphere Competence

The terminology associated with biocontrol in the rhizosphere and with soil-plant-microbe interactions has gradually become more complex through the use of a range of descriptive rather than mechanistic terms such as plant growth promotion and rhizosphere competence. Much like the situation with PGPR, many saprotrophic fungi, particularly certain isolates of Trichoderma species, can provide plant growth promotion in the absence of any major pathogens. In many cases these studies are restricted to simple observations of improved plant growth with no indication of the possible mechanisms involved, although there are exceptions. For example, Trichoderma harzianum 1295-27 was shown to solubilise phosphate and micro-nutrients that could be made available to provide plant growth.

This situation is compounded by the fact that many proven fungal biocontrol agents including some Trichodermaspecies, binucleate Rhizoctonia isolates and Pythium oligandrum can provide improved plant growth in the absence of pathogens. Further, colonisation of the surface of the seeds or roots or behaviour as endophytes has frequently been seen to be a desirable trait for biocontrol activity and although there is a clear relationship between rhizosphere colonisation and biocontrol activity with some isolates of biocontrol fungi such as Trichoderma species, non-pathogenic Fusaria, P. oligandrum,Verticillium biguttatum, and Talaromyces flavus, this is not always the case. Indeed transient plant growth inhibition following application of some biocontrol agents to seeds or roots is known. Consequently, it is important to appreciate that just because a micro-organism can grow in the rhizosphere or spermosphere, it may not automatically provide biocontrol or plant growth promotion. Similarly, the converse is true. A proven biocontrol agent of a soil-borne plant pathogen may not always be capable of colonising the rhizosphere or providing plant growth promotion. The situation is much clearer with mycorrhizal fungi where, through ecologically mutualistic symbiosis with the plant, the major feature involves improving plant nutritional status, perhaps water balance and thus plant growth. Biocontrol of plant pathogens is generally viewed as a secondary role. Interestingly, a non-mycorrhizal endophytic fungus Piriformospora indica Verma, Varma, Kost, Rexer and Franken has recently been shown to promote growth of a range of plant species and it would be valuable to understand the mechanisms of action.

The factors controlling rhizosphere competence are unclear. It has been suggested that ability to produce cellulases and thus utilise substrates available in the rhizosphere may be an important feature. However, UV mutants of Trichoderma harzianum lacking cellulase production were found to have enhanced rhizosphere competence whereas two cellulase overproducers were found not to colonise the rhizosphere of bean plants. In a different system, both pathogenic and non-pathogenic Fusarium oxysporum populations exhibited their own characteristic growth and development features, as well as nutritional competence, which were not related to the ability to grow in the rhizosphere or to infect roots of tomato. Factors controlling rhizosphere competence deserve further study.

Multiple Microbial Interactions

The majority of interactions considered so far concern a single pathogen and a single biocontrol agent in the rhizosphere. However, one way of improving biocontrol in the rhizosphere may be to add mixtures or combinations of biocontrol agents, particularly if they exhibit different or complementary modes of action or abilities to colonise root microsites. Such multiple interactions are the normal situation in the rhizosphere. Numerous permutations have been considered, including combinations of different bacteria, fungi and both bacteria and fungi. For example, a seed application of a combination of three PGPR, Bacillus pumilus Meyer and Gotheil, Bacillus subtilis (Ehrenberg) Cohn and Curtobacterium flaccumfaciens (Hedges) Collins and Jones provided greater control of several pathogens on cucumber (Cucumis sativa L.) than when any were inoculated singly, combinations of Paenibacillus sp. and a Streptomyces sp. suppressed Fusarium wilt of cucumber better than when either was used alone and a combination of Pseudomonas fluorescens and Stenotrophomonas maltophila improved protection of sugar beet against Pythium-mediated damping-off in comparison with either applied individually. Combinations of fungi and bacteria have also been shown to provide enhanced biocontrol. For instance, Trichoderma koningii Oud. combined with either Pseudomonas chlororaphis 30–84 or P. fluorescens Q2-87 provided greater suppression of take-all of wheat than T. koningiialone, Trichoderma (Gliocladium) virens GL-3 combined with Burkholderia cepacia (Burkholder) Yabunchi *et. al.* provided stands of pepper (Capsicum annuum L.) greater protection than either antagonist used alone in the presence of a mixture of up to four soil-borne pathogens and non-pathogenic Fusarium oxysporum

Fo47 combined with Pseudomonas putida WCS358 provided better suppression of Fusarium wilt of flax (Linum usitassimum L.) caused by F. oxysporum f. sp. lini (Bolley) Snyder and Hansen than either alone. In the latter study, a high population density of Fo47 was important for disease control in general. Only when Fo47 was present at low population density was enhanced activity of the combination seen. Enhanced plant growth promotion has also been recorded in the absence of pathogens by applications of combinations of bacterial or fungal plant growth promoting micro-organisms. For example, Trichoderma aureoviride Rifai inoculated with the arbuscular mycorrhizal fungus Glomus intraradices enhanced growth of Citrus reshni more than the G. intraradices used alone.

However, it is important when considering the use of mixtures or combinations of strains that no member of the mixture is inhibitory to another or interferes excessively with the existing, normal and non-pathogenic microbiota associated with the roots. Certainly there are examples of combinations of different bacteria or fungi providing no better or, in some cases, worse plant growth promotion or biocontrol than the isolates used singly. Similarly, a combination of Bacillus subtilis and non-pathogenic Fusarium oxysporum did not provide control of Fusarium wilt of chickpea (F. oxysporumi f. sp. ciceris) whereas either applied alone did. Several biocontrol agents including isolates of Pseudomonas, Gliocladium and Trichoderma species have been shown to have little or no adverse effect on establishment and function of arbuscular mycorrhizas although there are reports of adverse effects of some isolates of Trichoderma and Streptomyces griseoviridis Anderson *et. al.* on arbuscular mycorrhiza formation. Clearly, the complex interactions that can take place in the rhizosphere between biocontrol agents and the indigenous microbiota needs to be considered during development of commercial microbial products.

Chapter 20

Plant-soil Micro-organism Interactions

INTRODUCTION

Soil microbial communities can change in response to many different ecological factors. For example, associations with plants, including across plant productivity gradients, susceptibility to herbivores and even biological invasions have all been demonstrated to alter soil microbial communities. While several studies have documented the community response of microbes to plant functional groups and plant species, little is known about how soil microbial communities may respond to subtle gradients of plant genetic factors such as tree genotype in natural systems.

Feedbacks between plants and soils in natural systems have commonly been shown to influence soil microbial community composition or activity and even ecosystem processes, however most of this work has been focused at the level of plant functional group or species. An appreciation of the links between plant traits and belowground processes in terrestrial ecosystems suggests that individual plant genotypes could influence the associated belowground soil microbial community (and also *vice-versa*). The quantity and quality of plant litter inputs to the soil (both above- and belowground) influences substrate availability for soil micro-organisms and may control microbial community structure. Variation in the aboveground plant community (regardless of the mechanism), and subsequent changes in plant traits at the plant species and functional group scales can lead to shifts in microbial community structure.

For example, many studies have documented that soil microbial community composition or microbial biomass differs in soils beneath different dominant tree species. These differences are likely due to differences in leaf and root chemistry or production, root exudation or abiotic effects due to differences in canopy structure. Bartelt-Ryser *et. al.* (2005) found legacy effects of microbial community composition in soils after plants of different species were grown and then removed from plots suggesting that plant species can have long-term effects on microbial community composition even after the plant itself is gone. At the level of plant phenotype or genotype, short-term experiments have found variation in total PLFA as well as ectomycorrhizal colonisation between two clones but few differences in either arbuscular mycrorhizal fungi, bacterial communities or microbial biomass. These studies suggest that microbial dynamics may have the capacity to respond to plant traits at both inter- and intraspecific scales.

We hypothesised that after 13 years of leaf and root inputs to the soils beneath replicated Populus spp. genotypes in a common garden, plant genetic effects would result in distinct microbial biomass (*i.e.*, active biomass of fungi and bacteria in the soil) and microbial community compositions among replicate copies of each genotype. We tested this hypothesis within two Populus species and their naturally occurring hybrids. We considered microbial biomass nitrogen (N), total phospholipid fatty acid (PLFA) biomarkers and microbial community composition (measured by Bray-Curtis dissimilarity of PLFA biomarkers) to respond like a quantitative trait of their associated tree similar to other plant phenotypes displaying quantitative inheritance.

When these estimates are generated according to standard breeding designs (*e.g.*, among lineages of genotypes), quantitative genetic methods are appropriate for relating the observable components of phenotypic variance of the microbial community to the measurable components of plant genetic variance. We define community heritability as the tendency of replicated individuals of the same genotype to support the same biomass or community of micro-organisms in soils directly beneath that tree.

PLANT-SOIL INTERACTIONS

Plant growth and development largely depend on the combination and concentration of mineral nutrients available in the soil. Plants often face significant challenges in obtaining an adequate supply of these nutrients to meet the demands of basic cellular processes due to their relative immobility. A deficiency of any one of them may result in decreased plant productivity and/or fertility. Symptoms of nutrient deficiency may include stunted growth, death of plant tissue, or yellowing of the leaves caused by a reduced production of chlorophyll, a pigment needed for photosynthesis. Nutrient deficiency can have a significant impact on agriculture, resulting in reduced crop yield or reduced plant quality. Nutrient deficiency can also lead to reduced overall biodiversity since plants serve as the producers that support most food webs.

Changes in the climate and atmosphere can have serious effects on plants, including changes in the availability of certain nutrients. In a world of continual global climate change, it is important to understand the strategies that plants have evolved to allow them to cope with some of these obstacles.

Two classes of nutrients are considered essential for plants : macro-nutrients and micro-nutrients. Macro-nutrients are the building blocks of crucial cellular components like proteins and nucleic acids; as the name suggests, they are required in large quantities. Nitrogen, phosphorus, magnesium, and potassium are some of the most important macro-nutrients. Carbon, hydrogen, and oxygen are also considered macro-nutrients as they are required in large quantities to build the larger organic molecules of the cell; however, they represent the non-mineral class of macro-nutrients. Micro-nutrients, including iron, zinc, manganese, and copper, are required in very small amounts. Micro-nutrients are often required as cofactors for enzyme activity.

Mineral nutrients are usually obtained from the soil through plant roots, but many factors can affect the efficiency of nutrient acquisition. First, the chemistry and composition of certain soils can make it harder for plants to absorb nutrients. The nutrients may not be available in certain soils, or may be present in forms that the plants cannot use. Soil properties like water content, pH, and compaction may exacerbate these problems.

Second, some plants possess mechanisms or structural features that provide advantages when growing in certain types of nutrient limited soils. In fact, most plants have evolved nutrient uptake mechanisms that are adapted to their native soils and are initiated in an attempt to overcome nutrient limitations. One of the most universal adaptations to nutrient-limited soils is a change in root structure that may increase the overall surface area of the root to increase nutrient acquisition or may increase elongation of the root system to access new nutrient sources. These changes can lead to an increase in the allocation of resources to overall root growth, thus resulting in greater root to shoot ratios in nutrient-limited plants.

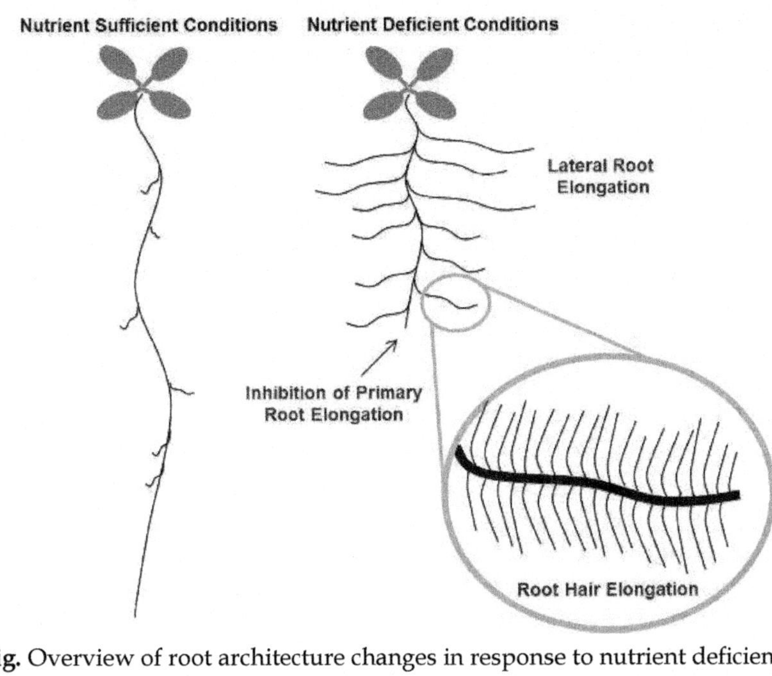

Fig. Overview of root architecture changes in response to nutrient deficiency.

Plants are known to show different responses to different specific nutrient deficiencies and the responses can vary between species. The most common changes are inhibition of primary root growth (often associated with P deficiency), increase in lateral root growth and density (often associated with N, P, Fe, and S deficiency) and increase in root hair growth and density (often associated with P and Fe deficiency).

Plant roots exhibit a variety of changes in response to nutrient deficiency, including inhibition of primary root elongation and increased growth and density of lateral roots and root hairs. These responses are species-, genotype-, and nutrient-specific, but they are generalised in this figure to demonstrate all potential effects.

While nutrient deficiencies can pose serious threats to plant productivity, nutrients can become toxic in excess, which is also problematic. When some micro-nutrients accumulate to very high levels in plants, they contribute to the generation of reactive oxygen species (ROS), which can cause extensive cellular damage. Some highly toxic elements like lead and cadmium cannot be distinguished from essential nutrients by the nutrient uptake systems in the plant root, which means that in contaminated soils, toxic elements may enter the food web via these nutrient uptake systems, causing reduced uptake of the essential nutrient and significantly reduced plant growth and quality.

In order to maintain nutrient homeostasis, plants must regulate nutrient uptake and must respond to changes in the soil as well as within the plant. Thus, plant species utilise various strategies for mobilisation and uptake of nutrients as well as chelation, transport between the various cells and organs of the plant and storage to achieve whole-plant nutrient homeostasis. Here, we briefly describe a few examples of strategies used by plants to acquire nutrients from the soil.

Plant Acquisition of Nutrients

Potassium. Potassium (K) is considered a macro-nutrient for plants and is the most abundant cation within plant cells. Potassium has a number of important functions within plants, including balancing the charges of cellular anions, enzyme activation, control of stomatal opening/closing and serving as an osmoticum for cellular growth.

Potassium deficiency occurs frequently in plants grown on sandy soils resulting in a number of symptoms including browning of leaves, curling of leaf tips and yellowing (chlorosis) of leaves, as well as reduced growth and fertility.

Potassium uptake processes have been the subject of intense study for several decades. Early studies indicated that plants utilise both high and low affinity transport systems to directly acquire potassium from the soil. Low affinity transport systems generally function when potassium levels in the soil are adequate for plant growth and development. This process is mediated by ion channels in the plasma membrane of root cells, allowing passive transport of K^+ from areas of relatively high external concentration into the plant cells where the concentration of K+ is lower. The expression of these low affinity transporters does not appear to be significantly affected by potassium availability.

Under conditions of potassium limitation, in contrast, plants usually induce high affinity K+ transport systems. There are likely many proteins involved in high affinity potassium transport, but in Arabidopsis, two proteins have been identified as the most important transporters in this process. Interestingly, one of these transporters, AtHAK5, is a carrier protein and thought to mediate active transport of potassium into plant roots, whereas the other protein, AKT1, is a channel protein and likely mediates a passive transport mechanism with an increased affinity for K+ under conditions of potassium limitation. More recent work shows that plants contain a number of different transport systems to acquire potassium from the soil and distribute it within the plants. Although much remains to be learned about potassium uptake and translocation in plants, it is clear that the mechanisms involved are complex and tightly controlled to allow the plant to acquire sufficient amounts of potassium from the soil under varying conditions.

Iron. Iron is essential for plant growth and development and is required as a cofactor for proteins that are involved in a number of important metabolic processes including photosynthesis and respiration. Despite the fact that iron is the fourth most abundant element in the earth's crust, it is often limiting for plants due to the fact that it tends to form insoluble complexes in aerobic soils of neutral to basic pH. It is thought that iron limitation is a problem for plants on as much as 30 per cent of soils worldwide. Iron-deficient plants often display interveinal chlorosis, in which the veins of the leaf remain green while the areas between the veins are yellow. Due to the limited solubility of iron in many soils, plants often must first mobilise iron in the rhizosphere (a region of the soil that surrounds, and is influenced by, the roots) before transporting it into the plant. Two distinct mechanisms have evolved that are utilised by plants to acquire iron from the soil, termed the Strategy I and Strategy II responses.

Fig. Iron-deficiency chlorosis in soybean.

The plant on the left is iron-deficient while the plant on the right is iron-sufficient.

Strategy I is used by all plants except the grasses. It is characterised by three major enzymatic activities that are induced in response to iron limitation and that are located at the plasma membrane of cells in the outer layer of the root. First, strategy I plants induce the activity of H+-ATPase, which uses the energy of ATP to pump protons out of the root cells and into the rhizosphere. This activity thus serves to acidify the rhizosphere and the drop in rhizosphere pH solubilises ferric iron (Fe3+), making it more available for subsequent uptake by the plant. Second, strategy I plants induce the activity of a plasma-membrane-bound ferric chelate reductase. Ferric chelate reductase activity reduces ferric iron to the more soluble ferrous form (Fe^{2+}) of iron. Finally, plants induce the activity of a ferrous iron transporter that moves ferrous iron across the plasma membrane and into the plant.

In contrast, the grasses utilise strategy II to acquire iron under conditions of iron limitation. Following the imposition of iron limitation, strategy II species begin to synthesise special molecules called phytosiderophores (PSs) that display high affinity for ferric iron. PSs are secreted into the rhizosphere where they bind tightly to ferric iron. Finally, the PS-ferric iron complexes are transported into root cells by PS-Fe(III) transporters. Interestingly, while both strategies are relatively effective at allowing plants to acquire iron from the soil, the strategy II response is thought to be more efficient because grass species tend to grow better in calcareous soils (which have a high pH and thus have limited iron available for uptake by plants).

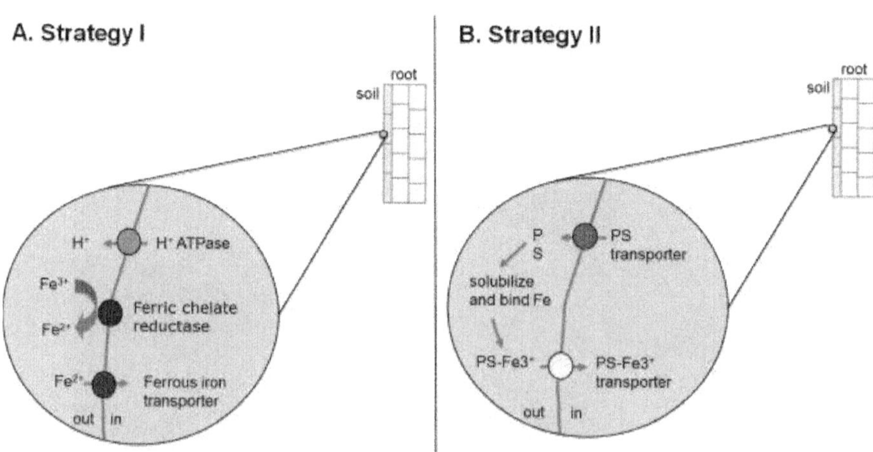

Fig. Strategy I and Strategy II Mechanisms for Iron Uptake.

A. Strategy I plants induce the activity of a proton ATPase, a ferric chelate reductase, and a ferrous iron transporter when faced with iron limitation. B. In contrast, Strategy II plants synthesise and secrete phytosiderophores (PS) into the soil in in response to iron deficiency. The PSs bind ferric iron with high affinity and the resulting PS-Fe3+ complexes are transported from the soil into the plant root.

Symbioses with Soil-based Micro-organisms

Nitrogen and phosphorus are among the elements considered most limiting to plant growth and productivity because they are often present in small quantities locally or are present in a form that cannot be used by the plant. As a result, the evolution of many plant species has included the development of mutually beneficial symbiotic relationships with soil-borne micro-organisms. In these relationships, both the host plant and the micro-organism symbiont derive valuable resources that they need for their own productivity and survival as a result of the association.

Nitrogen Fixation. Despite the fact that nitrogen is the most abundant gaseous element in the atmosphere, plants are unable to utilise the element in this form (N_2) and may experience nitrogen deficiency in some soils that have low nitrogen content. Since nitrogen is a primary component of both proteins and nucleic acids, nitrogen deficiency imposes significant limitations to plant productivity.

In an agricultural setting, nitrogen deficiency can be combated by the addition of nitrogen-rich fertilizers to increase the availability of nutrients and thereby increase crop yield. However, this can be a dangerous practice since excess nutrients generally end up in ground water, leading to eutrophication and subsequent oxygen deprivation of connected aquatic ecosystems.

Plants are able to directly acquire nitrate and ammonium from the soil. However, when these nitrogen sources are not available, certain species of plants from the family Fabaceae (legumes) initiate symbiotic relationships with a group of nitrogen fixing bacteria called Rhizobia.

These interactions are relatively specific and require that the host plant and the microbe recognise each other using chemical signals. The interaction begins when the plant releases compounds called flavanoids into the soil that attract the bacteria to the root. In response, the bacteria release compounds called Nod Factors (NF) that cause local changes in the structure of the root and root hairs. Specifically, the root hair curls sharply to envelop the bacteria in a small pocket. The plant cell wall is broken down and the plant cell membrane invaginates and forms a tunnel called an infection thread that grows to the cells of the root cortex.

The bacteria become wrapped in a plant derived membrane as they differentiate into structures called bacteroids. These structures are allowed to enter the cytoplasm of cortical cells where they convert atmospheric nitrogen to ammonia, a form that can be used by the plants. In return, the bacteroids receive photosynthetically derived carbohydrates to use for energy production (reviewed by Limpens and Bisseling, 2003; Ferguson *et. al.* 2010).

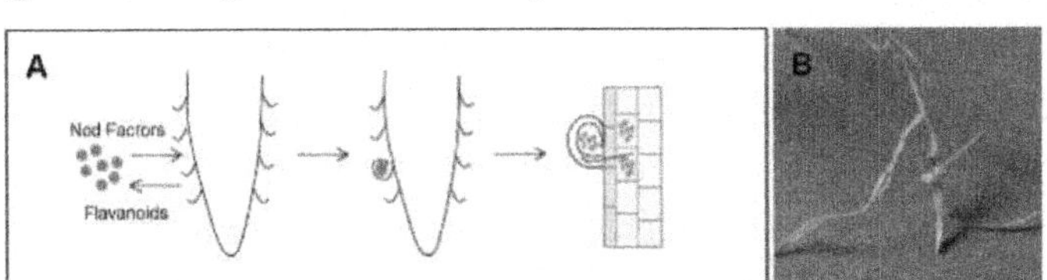

Fig. Nodulation of Legumes.

A. Process of root cell colonisation by rhizobacteria. B. Nodule formed by nitrogen fixing bacteria on a root of a pea plant (genus Pisum).

Mycorrhizal Interactions with Plants

In addition to symbiotic relationships with bacteria, plants can participate in symbiotic associations with fungal organisms as well. Current estimates of the frequency of plant-mycorrhizal associations suggest that around 80 per cent of all plants establish some type of mycorrhizal symbiosis, and many studies indicate that these relationships are millions of years old. There are several classes of mycorrhiza, differing in structural morphology, the method of colonising plant tissue, and the host plants colonised. However, there are two main classes that are generally regarded as the most common and therefore, the most ecologically significant. The endomycorrhizae are those fungi that establish associations with host plants by penetrating the cell wall of cortical cells in the plant roots. By contrast, ectomycorrizae develop a vast hyphae network between cortical cells but do not actually penetrate the cells.

The most common endomycorrhizal interaction occurs between arbuscular mycorrhizal fungi (AMF; also called Vesicular-Arbuscular Mycorrhiza or VAM) and a variety of species of grasses, herbs, trees and shrubs. When phosphate is available in the soil, plants are able to acquire it directly via root phosphate transporters. However, under low phosphate conditions, plants become reliant on interactions with mycorrhizal fungi for phosphorus acquisition. Mycorrhizal spores present in the soil are germinated by compounds released from the plant.

Hyphae extend from the germinating spore and penetrate the epidermis of the plant root. Inside the root, the hyphae branch and penetrate cortical cells, where highly branched structures called arbuscules develop. Externally, hyphae extend into the soil beyond the area accessible to the root.

This kind of symbiosis facilitates plant phosphorus uptake from the soil by increasing the root's absorptive surface area. Since plants take up phosphorus at a much higher rate than phosphorus diffuses into the soil surrounding the root, a phosphorus depletion zone is quickly established, limiting uptake of phosphorus by the plant. However, AMF hyphae form a bridge between the internal root environment and the area beyond this depletion zone enabling the plant to acquire significantly more phosphorus through its symbiotic partner than it could on its own.

A. Diagram of arbuscular mycorrhizae colonisation of a plant root showing the extension of hyphae beyond the phosphorus depletion zone and the presence of arbuscules in cells of the root cortex. B. Diagram of Ectomycorrhizal fungi showing growth of hyphae around cortical cells, a mantle sheath on the outside of the root, and hyphae that extend into soil around the root.

Ectomycorrhizal fungi (EcM) form associations with many tree species (birch, oak, spruce, pine, fir), making these micro-organisms an important component of the biodiversity found in forest ecosystems, particularly temperate and boreal forests in the northern hemisphere.

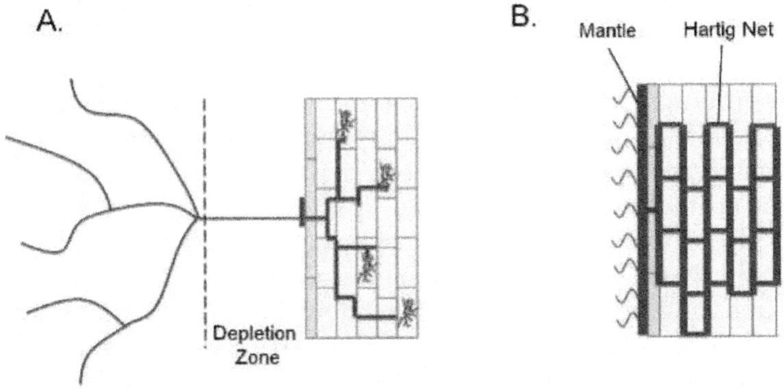

Fig. Plant-mycorrhizal Fungus Interactions.

Although the molecular mechanisms underlying initiation of EcM-plant interactions are not well defined, it is clear that signaling molecules are released by both symbiotic partners, resulting in fungal hyphae growth towards compatible plant roots. The EcM forms an extensive hyphal sheath, or mantle, around the root cap and a hartig net of hyphae that extends into the root itself and surrounds the cells within the root cortex. Hyphae from the mantle extend into the surrounding soil, providing availability and storage of nutrients like phosphorus and nitrogen for the plant. This is particularly important for trees growing in forest soils because a large portion of the nutrients available are located in the litter layer and are not accessible to the roots of these trees. The EcM produce enzymes that can digest this organic material and also have the ability to mobilise the nutrients to the hartig net, making them available to the plant. In this way, the interaction between trees and EcM is not only beneficial, but in many cases, is essential.

As is the case with bacterial symbioses, the benefit to the fungal partners in these relationships involves transfer of photosynthetic carbohydrates from the plant to the micro-organism for metabolism and energy production. However, the nutritional advantage for plants of such symbiotic relationships far outweighs the energetic cost imposed by hosting micro-organisms like rhizobacteria, AMF, and EcM.

METHODS OF SOIL MICROBIAL COMMUNITY

Utilising a common garden of known genotypes planted in 1991, we quantified the soil microbial community beneath three to five replicate clones each of multiple genotypes of Populus fremontii (four genotypes), P. angustifolia (five genotypes) and their natural F1 (five genotypes) and backcross hybrids (six genotypes) for a total of 70 trees. Tree genotypes were previously collected and replicated as stem cuttings from natural populations along a 105 km transect along the Weber River, northern UT, USA and planted randomly in a one hectare common garden in Ogden UT, on 5 meter centers.

Randomisation of replicate clones within the garden minimised site bias and maximised our ability to detect genetic effects. Soils at the garden were typical of low elevation alluvial riparian sites and were characterised as coarse-loamy, mixed, mesic Oxyaquic Haploxerolls. Due to 13 years of Populus litter accumulation and shading, the understory is sparse immediately adjacent to the tree trunk but includes Bromus tectorum, Ambrosia spp., Tragopogon dubious, Poa bulbusa, Lactuca serriola; the understory community showed no pattern of abundance relative to the tree genotypes or cross types (Kovacs *et. al.* pers. comm.). Tree genotypes and specific cross type status (*i.e.*, pure species, F1 or backcross hybrids) were previously determined with restriction fragment length polymorphism (RFLP) analyses. While genetic by environment interactions (G X E) are clearly important to microbial dynamics, common garden approaches to elucidate the role of the genetic vs. environmental components are valuable to help tease apart the genetic interactions that may tie plants to their associated communities in ways that are only beginning to be appreciated.

Microbial Analyses

In October 2004, when the trees were 13 years old (and 10-15 m tall), we sampled bulk, mineral soils beneath each tree using polycarbonate soil cores (4.8 cm X 15 cm). We chose this season as previous research in this system indicated that fall was an optimal time to quantify microbial activity due to increased rainfall and pulses of carbon from litterfall (as demonstrated by highest rates of net nitrogen mineralisation, relative to other seasons in the year; data from Schweitzer *et. al.* 2004). To minimise disturbance beneath each tree in this long-term common garden, soil core placement was standardised; a 15 cm deep soil core was placed and removed on the north side of each tree within 0.25 m of the trunk.

The soils were kept cool, sieved (< 2 mm) to remove coarse fragments and roots, and a sub-sample was immediately freeze-dried. From these samples, we assessed microbial community composition with PLFA biomarkers using the method of White and Ringleberg.

The freeze-dried soil was extracted with a phosphate-buffered chloroform-methanol solvent and separated into functional classes of signature fatty acids after methylation of the polar lipids and analysis by gas chromatography. The PLFAs are unique to major taxonomic groups. While this is a coarse measure of microbial community composition, recent evidence suggests it is just as effective in differentiating treatment patterns as functional analyses (*i.e.*, community-level physiological profiling) and molecular techniques. We conservatively used common PLFA between 14 and 18 carbon (C) atoms long to include in the analysis as those are PLFA exclusively identified as microbial. Total PLFA was determined as the sum of all PLFA biomarker concentrations. PLFA compounds identified as general bacterial and fungal biomarkers were used to calculate the ratio of bacterial to fungal PLFA concentration in the soils.

We also quantified microbial biomass N using the chloroform fumigation extraction method. A sub-sample of field moist soil (ca. 25 g) was extracted with 50 ml 0.5 M K2SO4, shaken, gravity-filtered with Whatman filter paper No. 1 (leached with deionised water and K2SO4) and stored in a freezer until chemical analysis. Another 25 g field moist sub-sample was exposed to chloroform for five days in a glass vacuum desiccator with 30 ml ethanol-free chloroform. After the fumigation period the samples were extracted with 50 ml of 0.5 M K2SO4 in the same manner as above. Total microbial biomass N was determined on the thawed extracts with a micro-Kjeldahl digestion followed by colourimetric analysis using the salicylate method on a Lachat AE auto-analyser.

Microbial biomass N was determined by subtracting the un-fumigated sample from the fumigated sample. A third sub-sample of field moist soil was oven dried (48 h at 105° C) to determine soil moisture content. All PLFA and microbial biomass values are expressed on an oven-dry mass basis.

Statistical Analyses

To detect differences in microbial biomass and microbial communities (based on the NMDS scores) among tree cross types, we performed a line-cross analysis of the size of the microbial biomass and the composition of the microbial community across the four cross types (Populus fremontii, P. angustifolia, and their natural F1 and backcross hybrids). A line-cross analysis estimates the additive (*i.e.*, parent to offspring genetic effects), dominant (*i.e.*, non-additive, one allele effects), and epistatic (*i.e.*, effects of non-additive novel combinations of genes) components of phenotypic variance associated with plant cross type and microbial dynamics across the hybridising complex. This allowed us to make inferences on the relative contribution of plant genotype *versus* hybrid cross type to microbial dynamics as well as how genetic introgression may affect microbial community dynamics.

We assessed plant genetic effects on microbial communities across the hybridising complex following the methods of Shuster *et. al.* We calculated a dissimilarity matrix of the PLFA from soils collected beneath individual tree genotypes using the Bray-Curtis dissimilarity coefficient. The major directions of microbial compositional change were summarised and quantified using non-metric multi-dimensional scaling (NMDS) with a one-dimensional solution. We extracted the NMDS score which represents an estimate of the combined PLFA richness (*i.e.*, functional classes) and abundance (based on nmol PLFA/g of dry soil) for each replicated clone of each tree genotype.

To analyse differences in microbial biomass and microbial community composition among genotypes within each cross type, we performed a univariate ANOVA of microbial data within each cross type using JMP 5.1 (SAS Institute Cary, NC, USA). The data showed no evidence for violation of assumptions of normality and homogeneity of variance. From the variance component estimates from the ANOVA analyses, we calculated the broad-sense heritability (*i.e.*, clonal repeatability) of microbial biomass (H^2B ± 95 per cent confidence limits) and microbial community composition (H^2C ± 95 per cent confidence limits) as the among-genotype component of variation in the community phenotype (s^2_S) divided by the total variance in the community phenotype for all replicate trees (s^2_{total}), where the total variance in the phenotype was the sum of the among genotype and the error component. To determine if particular PLFA were structuring the microbial community (as opposed to a whole community response), we used Indicator Species Analysis using PC-ORD 4.02.

This analysis calculates an indicator value based on fidelity and relative abundance of species (in this case, PLFA) to each group (in this case plant genotype). Indicator species (those with significant Indicator Values) are defined as those expressing the highest degree of specificity and fidelity to a group or site, independent of abundance, following this basic formula :

$$A_{ij} \times B_{ij} \times 100 = IndVal_{ij} \text{ and } \max [IndVal_{ij}] = IndVal_i$$

A_{ij} is a specificity measure (A_{ij} = Nindividualsij/Nindividualsi) and B_{ij} is a fidelity measure. A significant indicator species for a particular site or group was determined by a Monte Carlo test based on Bray Curtis distance.

Microbial Biomass

The line-cross analyses used to estimate the plant genetic contributions to microbial dynamics across the hybridising complex show that the model incorporating additive genetic effects was not rejected (c^2=5.98 and c^2=0.05; df=2, for total PLFA and microbial biomass N, respectively).

However, the additive genetic coefficient was not significant, indicating that variation in microbial biomass within cross types overwhelms the variation among tree cross types. In other words, plant genotype explained an overwhelming proportion of the variance in total PLFA and microbial biomass N relative to the differences between the species and their naturally occurring hybrids.

Consistent with this observation, we found significant differences in microbial biomass N within cross types. Microbial biomass N differs among genotypes within P. angustifolia ($F_{4,14} = 4.95$, $p = 0.026$), as well as the F_1 ($F_{4,16} = 6.01$, $p = 0.007$) and backcross hybrids ($F_{5,19} = 3.67$, $p = 0.025$). While there was variation, there were no significant difference in microbial biomass N among genotypes within P. fremontii ($F_{3,13} = 0.28$, $p = 0.84$). From the variance estimates for microbial biomass N, we found significant broad-sense heritability among genotypes within P. angustifolia ($H^2B = 0.61 \pm 0.59$; [95 per cent confidence intervals]), the F_1 hybrids ($H^2B = 0.23 \pm 0.09$) and the backcross hybrids.

Similarly, for total PLFA (another estimate of microbial biomass), we found significant differences among genotypes in P. angustifolia ($F_{4,14} = 15.57$, $p = 0.0002$) and the backcross hybrids ($F_{5,19} = 3.25$, $p = 0.04$). There were no differences among genotypes for P. fremontii or the F1 hybrids. The variance estimates for total PLFA suggest that plant genotype explained 78 per cent ($H^2B = 0.78 \pm 0.032$) of the variation within P. angustifolia genotypes and 43 per cent of the variation within backcross hybrids ($H^2B = 0.43 \pm 0.43$, although this is not significant due to overlapping of confidence intervals).

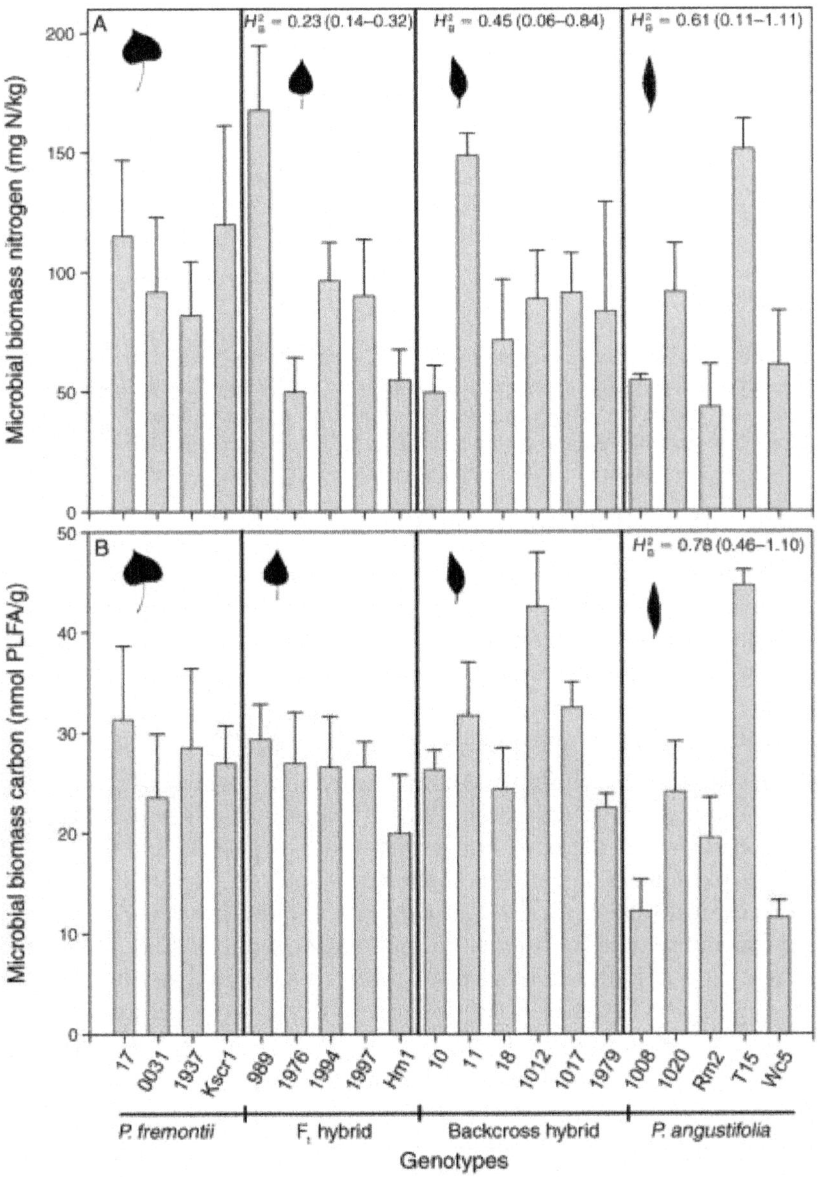

Fig. Mean (± 1 SE) microbial biomass nitrogen (N) (A) and total phospholipids fatty acid (PLFA) biomarkers (B) beneath replicated genotypes of P. fremontii, F1 hybrids, backcross hybrids and Populus angustifolia in a common garden. Broad-sense heritabilities (H^2B ± 95 per cent confidence limits) are listed where significant.

These results indicate that tree genotypes within species (*i.e.*, intraspecific genetic variation) can create environments to which soil micro-organisms consistently respond, leading to potential feedbacks in nutrient availability and C storage in soils beneath individual trees.

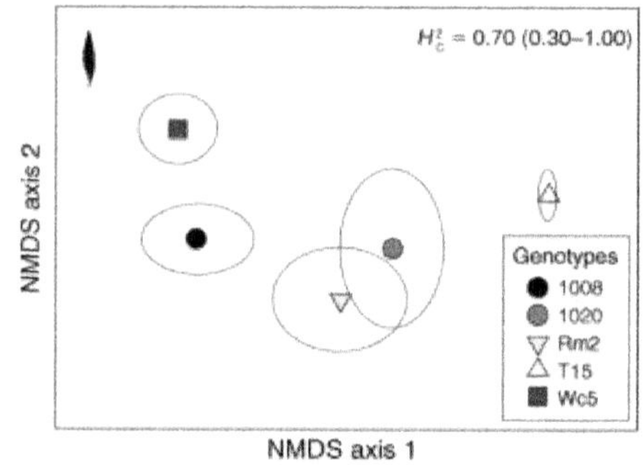

Fig. Mean microbial community composition (*i.e.*, Bray-Curtis dissimilarity) of an individual plant genotype (surrounded by ellipses representing ± 1 SE; n = 3-5 replicates per genotype) underneath replicated tree genotypes of Populus angustifolia in a common garden. As this is a distance measure, the axes are unit-less. H^2C (± 95 per cent confidence limits) represents the broad-sense heritability for the microbial community composition.

Microbial Community Composition

A line-cross analysis revealed that additive (χ^2=9.79) and additive plus dominance models (χ^2=8.63) were rejected (df = 2; data not shown), suggesting that variation in microbial community composition within tree cross types was higher than differences among tree cross types. These results also suggest that epistatic genetic interactions across the hybridising complex likely have a large effect on the microbial community dynamics among the cross types.

When we examined microbial community composition of genotypes within each cross type, we only found significant differences in microbial community composition among P. angustifolia genotypes ($F_{4,14}$ = 8.31, p = 0.0002). Plant genetic factors explained 70 per cent of the variation (*i.e.*, H^2C = 0.70 ± 0.40) in soil microbial communities within P. angustifolia. Furthermore we found no significant patterns in the types of micro-organisms associated with the tree cross types, indicating that our patterns were not driven by specific categories or "blooms" of micro-organisms in the soil. There were no differences within any of the cross types in the ratio of bacterial to fungal PLFA among genotypes (data not shown). Similarly, when we examined the effects of plant genetic factors on individual PLFA biomarkers we only found patterns within backcross hybrids, suggesting that in the pure species and F1 hybrids, no single PLFA was specific to those tree types. PLFA i15 : 0, a15 : 0, 17 : 0, cy17 : 0 and 18 : 2w6t were significant indicators for backcross hybrids, suggesting strong associations of the soil micro-organisms containing these biomarkers with backcross tree genotypes alone. However as there were no overall significant differences in community composition between genotypes in backcross hybrids, the meaning of this result is unclear.

Discussion

After 13 years of plants conditioning the soils beneath them, we found that intraspecific plant genetic variation can affect soil microbial biomass and microbial community composition. These data support the hypothesis that feedbacks from plant genotype may be an important ecological factor affecting microbial dynamics. Overall, we found that plant genotype influenced microbial biomass N beneath replicated genotypes of P. angustifolia, F1 and backcross hybrids (but not P. fremontii genotypes) and demonstrated significant broadsense heritability.

Plant genetic factors explained 61 per cent, 23 per cent and 41 per cent of the variation in microbial biomass N, respectively. Total PLFA, as another measure of microbial biomass, demonstrated significant genotype effects in soils associated with P. angustifolia and backcross hybrid genotypes. Genotype was a significant predictor of soil microbial community composition within genotypes of P. angustifolia, with plant genotype explaining 70 per cent of the variation in community composition.

Among cross types (excepting backcross hybrids) we found no specific PLFA were associated or specific to tree cross type; the bacteria to fungi ratio also did not differ among genotypes in any of the tree cross types. The lack of significant indicators suggests that the observed patterns were not driven by any particular microbial group. In total these results demonstrate that microbial biomass N and sometimes microbial community composition vary by plant genotype, which suggests that plant genotype can select for particular soil microbial characteristics.

Genotypic Variation

While the studies to date in this system have focused on the pure and hybrid cross type level, only recently have we begun to investigate the within class variation. For example, when plant genotype was examined, genotype explained a greater proportion of the variance in arthropod community composition than species or hybrid cross type.

Moreover, the patterns of broad-sense heritability of arthropod community composition followed a similar gradient with the highest H2C occurring in the P. angustifolia and backcross hybrids and the lowest H2C in the P. fremontii. The consistency of these two studies suggest that there are similar patterns of genetic variation within the pure and hybrid cross types (*i.e.*, low genetic variation in P. fremontii and high genetic variation in P. angustifolia), which is consistent with past research.

Results from these two studies indicate that, regardless of the response variable (*i.e.*, arthropod communities - see Shuster *et. al.* 2006, microbial biomass N, and microbial community composition - this study), intraspecific plant genetic variation explains a greater proportion of the variance than do the differences between species and their hybrids. From an evolutionary perspective this should not be unexpected.

Populations of genotypes that are distributed across the landscape face a variety of biotic and abiotic selective forces which results in genotypes with unique suites of traits, genetic architecture, and therefore unique evolutionary dynamics. Other studies of genotype-specific responses in many response variables (*e.g.*

arthropod communities, trophic-level interactions and litter decomposition) are being demonstrated across plant species and suggest that there may be much genetic variation (for any given trait) within species.

Demonstration of broad-sense heritability, or clonal repeatability, in microbial biomass and community composition (at least in P. angustifolia) suggests that tree genotypes within species (*i.e.*, intraspecific genetic variation) can create environments to which soil micro-organisms may respond. However in natural systems, experiments at the genotype level examining plant-microbial interactions have been inconclusive, mostly due to lack of replication.

The studies to date also suggest that belowground components (*i.e.*, roots and/or root exudates) may play a large role in structuring the soil microbial community. For example, Bever *et. al.* found no effect of plant genotype on arbuscular mycorrhizal communities in a greenhouse study, although the study only utilised one genotype each of four different plant species. Similarly, Madritch and Hunter found that after 24 and 36 months, respectively, of litter additions by individual oak phenotypes into experimental microcosms, there was no effect of litter phenotype on either soil microbial biomass or microbial community composition.

A study that examined the effects of both above- and belowground inputs (*i.e.*, leaf litter and roots) across replicates of two clones found significant clonal effects on total PLFA and on ectomycorrhizal colonisation. The latter studies support the idea that clonal variation in belowground properties may impact soil microbial characteristics, which is supported by our data. Similarly, in agricultural systems, variation in the soil microbial community size or composition has been correlated with plant cultivar. Using clonal replicates of 20 plant genotypes we found that not only is clonal variation important to some microbial characteristics but that sometimes there is predicable variation in the community signature.

Heritability of Microbial Dynamics

Heritability of community composition was recently empirically demonstrated and mathematically defined as H^2C in Shuster *et. al.* Based on theory outlined by Shuster *et. al.*, the tendency of replicated individuals of the same genotype to support the same biomass or community of micro-organisms in soils directly beneath that tree (*i.e.*, community heritability) suggests that there are correlated fitness effects arising from interspecific indirect genetic effects (IIGEs) between the associated host plant and dependent interacting species in a community. However, just like traditional studies of phenotypic heritability, the amount of variance explained by plant genetic factors may be largely context dependent.

Inter-specific indirect genetic effects provide the theoretical basis for individual plant genotypes having heritable community phenotypes such as what has been observed with arthropod communities and now some microbial characteristics. For example, Shuster *et. al.* found significant heritable variation in arthropod communities within P. fremontii, P. angustifolia and their backcross hybrids. Similarly, here we found significant heritable variation in microbial community composition, but only within P. angustifolia. These data suggest that there may be feedbacks at the genotype level to associated soil micro-organisms (and potentially back to the plants), even though our test of microbial community composition (PLFA) may have been too coarse to detect differences that may actually exist.

IIGEs account for the interactions of the whole community (*i.e.*, the plant and each microbial community member) in which fitness of individual members is affected by their interactions within the community. While one may be able to identify all of the interacting phenotypes and their fitness consequences within a simple two or three species community, identifying all of the interacting phenotypes in a complex community is just not tractable, particularly in a microbial system.

Therefore all that can be said at present is that IIGE represent the likely mechanism for the heritable variation in P. angustifolia genotypes microbial community composition patterns.

Above- and Belowground Mechanisms

Our demonstration of broad-sense heritability, or clonal repeatability, in microbial biomass and community composition demonstrates that soil micro-organisms can consistently respond (within a genotype) to the environment related to intraspecific genetic variation. Whether the plant genetic factors affecting microbial dynamics are related to above- or belowground processes is currently unknown. Previous work in this Populus system have shown that tree cross types (and genotypes) differ in many traits that could impact associated soil micro-organisms.

For example, plant secondary chemistry (condensed tannins, phenolics) and nutrients, differ between the foliage of different cross types. P. fremontii and F1 hybrids typically demonstrate low concentrations of condensed tannin while the backcross hybrids and P. angustifolia genotypes generally have high concentrations of condensed tannin in their leaf, twig and bark tissues.

Furthermore the concentration of foliar condensed tannin demonstrates broad-sense heritability within P. angustifolia. Similarly, fine root production, soil CO_2 efflux and total belowground carbon allocation and aboveground net primary productivity differ across the hybridising complex and demonstrate broad-sense heritability in aboveground net primary productivity.

Trees producing high condensed tannin in their tissues have higher fine root biomass but lower overall TBCA and AGNPP than trees with low condensed tannins in their tissues. It is likely a combination of these genetic-based above- and belowground trait differences that result in specific microbial phenotypes across replicate clones, especially with respect to microbial biomass. As Madritch and Hunter showed no effect of a 3 years litter phenotype manipulation, and a similar short-term litter manipulation experiment in the Populus system also show no effect of litter cross type (Fisher unpublished data), variation in belowground traits could be responsible for the patterns in microbial dynamics that we documented.

INTERACTIONS AMONG SOIL MICRO-ORGANISMS

Soil is the largest terrestrial ecosystem where a wide variety of relationships exists between different types of soil organisms. The associations existing between different soil micro-organisms, whether of a symbiotic or antagonistic nature, influence the activities of micro-organisms in the soil. Microflora composition of any habitat is governed by the biological equilibrium created by the associations and interactions of all individuals found in the community. In soil, many micro-organisms live in close proximity and interact among themselves in a different ways. Some of the interactions or associations are mutually beneficial, or mutually detrimental or neutral.

The various types of possible interactions/associations occurring among the micro-organisms in soil can be :

- Beneficial
  - Mutualism
  - Commensalisms and
  - Proto-cooperation or
- Detrimental/ harmful -
  - Amensalism,
  - Antagonism,
  - Competition
  - Parasitism and
  - Predation.

Beneficial Association/Interactions

- *Mutualism (Symbiosis) :* It is a relationship or a type of symbiosis in which both the interacting organisms/partners are benefited from each other. The way/manner in which benefit is derived depends on the type of interactions. When the benefit is in the term of exchange of nutrients, then the relationship is termed as "syntrophism" (Greek meaning : Syn-mutual and trophe = nourishment), for example, Lichen (association of algae or BGA with fungus) in which algae benefits by protection afforded to it by the fungal hyphae from environmental stresses, while the fungus obtain and use CO_2 released by the algae during photosynthesis. Where the blue green algae are the partners in the lichen association, the heterotrophs (Fungus), benefit from the fixed nitrogen by the blue green algae.

Micro-organisms may also form mutualistic relationships with plants, for example nitrogen fixing bacteria *i.e.* Rhizobium growing in the roots of legumes. In this Rhizobium-legume association, Rhizobium bacteria are benefited by protection from the environmental stresses while in turn plant is benefited by getting readily available nitrate nitrogen released by the bacterial partner.

The Anabaena-Azolla is an association between the water fern Azolla and the cyanobacterium Amabaena. This association is of great importance in paddy fields, where nitrogen is frequently a limiting nutrient.

An actinorrhizal symbiosis of actinomycetes, Frankia with the roots of Alnus and Casurina (non-legumes) is common in temperate forest ecosystem for soil nitrogen economy. Another type of symbiotic association which exists between the roots of higher plants and fungus is Mycorrhiza. In this association fungus gets essential organic nutrients and protection form roots of the plants and allows them to multiply and in turn plants uptake phosphorus, nitrogen and other inorganic nutrients made available by the fungus.

- *Commensalisms :* In this association one organism/ partner in association is benefited by other partner without affecting it. For example, many fungi can degrade cellulose to glucose, which is utilised by many bacteria. Lignin which is major constituent of woody plants and is usually resistant to degradation by most of the micro-organisms but in forest soils, lignin is readily degraded by a group of Basidiomycetous fungi and the degraded products are used by several other fungi and bacteria which can not utilise lignin directly. This type of association is also found in organic matter decomposition process.
- *Proto-cooperation :* It is mutually beneficial association between two species/ partners. Unlike symbiosis, proto-co-operation is not obligatory for their existence or performance of a particular activity. In this type of association one organism favour its associate by removing toxic substances from the habitat and simultaneously obtain carbon products made by the another associate/partner. Nutritional proto-co-operation between bacteria and fungi has been reported for various vitamins, amino and purines in terrestrial ecosystem and are very useful in agriculture. Proto-co-operative associations found beneficial in agriculture are :
  - Synergism between VAM fungus-legume plants andRhizobium in which nitrogen fixation and phosphorus availability/ uptake is much higher resulting in higher crop yields and improved soil fertility,
  - Synergism between PSM-legume plants and Rhizobium and
  - Synergism between plant roots and PGPR in rhizosphere where rhizobacteria restrict the growth of phytopathogens on plant roots and secrets growth promoting substances.

Detrimental (Harmful) Associations/Interactions

- *Antagonism :* It is the relationship in which one species of an organism is inhibited or adversely affected by another species in the same environment. In such antagonism, one organism may directly or indirectly inhibit the activities of the other. Antagonistic relations are most common in nature and are also important for the production of antibiotics. The phenomenon of antagonism may be categorised into three *i.e.* antibiosis, competition and exploitation.

In the process of antibiosis, the antibiotics or metabolites produced by one organism inhibits another organism. An antibiotic is a microbial inhibitor of biological origin. Innumerable examples of antibiosis are found in soil. For example, Bacillus Species from soil produces an antifungal agent which inhibits growth of several soil fungi. Several species ofStreptomyces from soil produces antibacterial and antifungal antibiotics. Most of the commercial antibiotics such as streptomycin, chloramphenicol, Terramycin and cyclohexamide have been produced from the mass culture of Streptomyces. Thus, species of Streptomyces are the largest group of antibiotic producer's in soil. Another example of antibiosis is inhibition of Verticillium by Trichoderma, inhibition of Rhizoctonia by a bacterium Bacillus subtilis, inhibition of soil fungus Aspergillus terreus by a bacterium Staphylococcus aureus.

- *Ammensalism :* In this interaction/association one partner suppress the growth of other partner by producing toxins like antibiotics and harmful gases like ethylene, HCN, Nitrite etc.
- *Competition :* As soil, is inhabited by many different species of micro-organisms, there exists an active competition among them for available nutrients and space. The limiting substrate may result in favouring one species over another. Thus, competition can be defined as "the injurious effect of one organism on another because of the removal of some resource of the environment". This phenomenon can result in major fluctuations in the composition of the microbial population in the soil.

For example, chlamydospores of Fusarium, Oospores of Aphanomyces and conidia of Verticillium dahlae require exogenous nutrients to germinate in soil. But other fungi and soil bacteria deplete these critical nutrients required for spore germination and thereby hinder the spore germination resulting into the decrease in population. Competition for free space has been, reported to suppress the fungal population by soil bacteria. Therefore, organisms with inherent ability to grow fast are better competitors.

- *Parasitism :* It is an association, in which one organism lives in or on the body of another. The parasite is dependent upon the host and lives in intimate physical contact and forms metabolic association with the host. So this is a host-parasite relationship in which one (parasite) is benefited while other (host) is adversely affected, although not necessarily killed. Parasitism is widely spread in soil communities, for example, bacteriophages (viruses which attack bacteria) are strict intracellular parasites Chytrid fungi, which parasitise algae, as well as other fungi and plants; there are many strains of fungi which are parasitic on algae, plants, animals parasitised by different organisms, earthworms are parasitised by fungi, bacteria, viruses etc.
- *Predation :* Predation is an association/ exploitation in which predator organism directly feed on and kills the pray organism. It is one of the most dramatic inter relationship among micro-organisms in nature, for example, the nematophagous fungi are the best examples of predatory soil

fungi. Species of Arthrobotrytis and Dactylella are known as nematode trapping fungi. Other examples of microbial predators are the protozoa and slime mold fungi which feed on the bacteria and reduce their population. The bacteriophages may also be considered as predators of bacteria.

Chapter 21

Biological Control of Rhizoctonia Solani in Soil

INTRODUCTION

Rhizoctonia solani Kühn is a widespread soil-borne pathogen that causes economically important diseases in many crops. Rice sheath blight caused by R. solani is one of the most important diseases of rice worldwide, causing considerable yield losses. The widespread adoptions of new, susceptible, high-yielding cultivars, which have large numbers of tillers, and the changes in cultural practices associated with these cultivars favour the development of sheath blight and contribute greatly to its rapid increase in incidence and severity in rice-producing areas throughout the world. Furthermore, environmental conditions such as low light, cloudy days, high temperature and high relative humidity also favour the disease development. The pathogen overwinters as soil-borne sclerotia and mycelium in plant debris; these constitute the primary inoculum. Control of the pathogen is difficult because of its ecological behaviour, its extremely broad host range and the high survival rate of sclerotia under various environmental conditions. So far, no rice variety completely resistant to this fungus has been found, although extensive evaluation of rice germplasm has been conducted. In the absence of desired level of host resistance, the disease is currently managed through excessive application of chemical fungicides, which have drastic effects on the soil biota, pollute the atmosphere, and are environmentally harmful. Some potentially effective fungicides are highly phytotoxic to rice and if the disease is not severe, these fungicides may reduce yields. Since it is difficult to achieve control through host resistance or fungicides, biological control may be effective in minimising the incidence of sheath blight.

The anamorphic fungal genus Trichoderma contains cosmopolitan soil-inhabiting fungi that are the major component of the mycoflora in soils of various ecosystems. The genus Trichoderma is especially known for its antagonistic activities against several plant pathogens including R. solani, and some strains are already commercialised as biocontrol agents (BCAs). They are potential agents in suppressing rice sheath blight and are highly competitive for rice plant residue and thus exhaust the nutrient supply for the pathogen and therefore greatly reduce its survival.

Sheath blight is the most important disease of high yielding rice cultivars in Mazandaran, the largest rice growing province in Iran, which is a Mediterranean region at the Southern coast of the Caspian Sea. The objective of this study was to evaluate the potential of indigenous Trichoderma isolates recovered from paddy rice fields in controlling R. solani, the rice sheath blight pathogen in vitro and in vivo.

IN VITRO RHIZOCTONIA SOLANI GROWTH INHIBITION ASSAYS

The methodology of Carruthers *et. al.*, with modifications, was used to determine fungal growth inhibition capacity of P. fluorescens isolates. Bacterial colonies grown for 48 h were streaked on the edges of Potato PDA plates and incubated at 25°C for 72 h. A 0.5 cm2 plug of a six days old R. solani culture was inoculated in the middle of the plate. Finally the plates were incubated at 25° C for six days and scored for inhibition of fungal growth by measuring the halo of growth daily in millimetres. The fungi inoculated with a non-antagonistic strain of Escherichia coli were used as negative control. This assay was done as a Completely Random Design (CRD), having five replicas and two repetitions. The isolates used for this test were : IBPF.25, IBPF.29, IBPF.30, IBPF.33, IBPF.60, IBPF.62, IBPF.63, IBPF.66, IBPF.78, IBPF.96 and IBPF.102. The results were analysed using SAS's ANOVA with SAS PROC GLM.

Pseudomonas fluorescens Antagonism Against R. Solani Under Greenhouse Conditions

Six P. fluorescens strains characterised by presenting none, moderate and high fungal growth inhibition *in vitro*, were selected for the evaluation in greenhouse conditions. All strains showed negative potential for the potato tissue maceration test. This test was done to eliminate the risk of working with any isolate capable of causing potato rot under storage conditions.

Colonies with natural resistance to rifampicin were selected in order to be able to identify thePseudomonas strains of our interest and distinguish them from the natural population present in the soil. Selected strains presented total sensitivity at concentrations of 25 μg ml1 or higher. No changes, in growth curves, colony morphology, duplication time, fluorescence pattern or *in vitro* antagonism, were found after the antibiotic selection using a concentration of 10 μg ml1.

The methodology reported by van Peer *et. al.*, and Kloepper *et. al.*, (1980), was used for bacterial suspension preparation. A combined treatment maintaining the same amount of cfu ml1 (108 cfu ml1), was prepared by mixing equal volumes of the bacterial suspension of each isolate. The production of R. solani and T. hamatum was carried out by inoculation in sterile wheat grains (previously soaked for

24 h in tap water), with 1 cm2 piece of mycelia grown in PDA. The wheat grains were then incubated at 25°C for three weeks.

Rhizoctonia solani and T. hamatum inoculums were added to soil as previously described by De Freitas and Germida. Trychoderma hamatum strain was added to the soil eight days before planting the potato seeds, whilst the pathogen was inoculated one day before. Plastic plantpots, containing approximately 2 kg of sterilised soil (soil texture sandy clay loam (sand 56 per centsilt 9.12 per centclay 34.9 per cent), pH : 5.0), were used to carry out root colonisation and antagonism tests. The soil was sterilised in extra-thick cardboard boxes, in two cycles of one hour at 121°C and 15 psi. Tuberseeds of S. phureja variety "clone one" were inoculated by submersion in the bacterial suspension. The untreated control and the treatment with R. solani alone were submerged in sterile 0.1 M MgSO4.

A CRD having seven replicas, was used for the evaluation of the seven treatments (six P. fluorescens isolates and the mix of isolates), against R. solani. A negative control (with neither pathogen nor antagonist), a disease marker (pathogen without antagonist) and a positive biological control (T. hamatum T21) were used. Disease presence was determined by the evaluation of tuber deformation, sclerotia and scarification formation in the tuber. Each symptom was divided in three categories as proposed by Delgado and Vargas : deformed tubers : was divided in tuber without damage, moderate deformation (having depressions) and advanced deformation (abnormal lumps and growth). Sclerotia and scarification were divided in healt, moderate (less than 40 per cent of tuber surface affection) and severe (more than 40 per cent of tuber surface affection).

Root Colonisation

The rhizospheric fluorescent Pseudomonas population resistant to rifampicin (10 μg ml1) was determined as cfu g1of root in each one of these samples, following the methodology proposed by Kloepper *et. al.* and Lemanceau *et. al.* The dynamics of root colonisation of the isolates was proposed as a CRD with seven treatments using a control (inoculated with 0.1 M MgSO4).

Three replicas and five sampling points each 2225 d, were analysed per treatment. A CRD having 2x8 factorial structure, with 16 treatments (six isolates, the bacterial mix and the control with and without R. solani,) and three replicas, was carried out to determine the possible influence of R. solani on P. fluorescens isolates colonisation. The values were collected at the end of the culture cycle (day 122) and transformed logarithmically (log 10) for the root colonisation analysis.

Plant Production and Growth Promotion

Plant productivity was evaluated after harvesting by taking the number and average weight of tubers produced per plant. At this point, foliage height (in cm) and dry weight (in g) were also determined for plant growth promotion in the presence and absence of pathogen.

The dry weight was determined by leaving the fresh plants in an oven at a temperate room at 40°C during eight days. These assays were carried out as a CRD having a 2x8 factorial structure, with 16 treatments and seven replicas. The T. hamatum isolate (T21) was used as a positive biological control.

Statistical Analysis

Root colonisation and plant growth during culture cycle were carried out using growth measurements that were analysed with SAS's PROC GLM. Other assays were analysed using one and two ways ANOVA analysis, with SAS's PROC GLM.

Pseudomonas Fluorescens Antagonist Effectivity Against R. Solani in Greenhouse Conditions

All tested strains of P. fluorescens significantly reduced ($p=0.0001^*$) the severity of the disease in the plants grown in greenhouse conditions. The tubers collected from all the plants inoculated with Pseudomonas isolates presented at least 50 per cent less incidence in all the symptoms at the moderate level. Pseudomonas fluorescens application also prevented severe levels of disease by comparison with the disease marker, in which severe deformation, sclerotia and scarification were produced in 9.03 per cent, 9.34 per cent and 6.54 per cent of the tubers respectively. In the same way, T. hamatum T21 application significantly reduced the symptoms of the disease and did not allow severe levels to be expressed.

Tuber Yield in the Presence and Absence of R. Solani

Pseudomonas fluorescens inoculation into S. phureja plants in the absence of the pathogen, increased tuber number and weight significantly in comparison with the untreated control and the disease marker. The IBPf.63 P. fluorescens strain and T. hamatum T21 positive control presented the greatest yield increase, with around two fold in comparison with both, the untreated control and the disease marker. The other strains although had lower tuber production increases (between 25 per cent and 48 per cent) and tuber weight gain (between 11 per cent and 74 per cent), were significantly greater than the control. It is important to mention that despite the differences in tuber number and weight between strains, the presence of R. solani did not affect those values within treatments ($p=0.077$). In the presence of the pathogen, the inoculation of P. fluorescens and T. hamatum T21, avoid the development of deformed small or large tubers, which were present in the disease marker (results not shown). This probably explains why a similar weight value in the disease marker, to that reached by tubers exposed to thePseudomonas strains ($p=0.506$), was obtained.

Solanum Phureja Growth Promotion Due to the Action of P. Fluorescens

The height of the plants inoculated with P. fluorescens was superior to that obtained by control plants (non-inoculated) from day 72 onwards (results not shown). In the last stage of the cycle (day 122), P. fluorescens inoculants produced an increase in height of at least 11 per cent and 40 per cent in dry weight, in relation to the control. The isolates IBPF.63 and IBPF.96 produced the greatest increases, equivalent to 71 per cent and 59 per cent in height and 110 per cent and 90 per cent in dry weight, respectively (p=0.0001). The growth promotion of the foliar area in the presence of the pathogen was also shown by the isolates IBPf.63 and IBPf.96 with increases of 128 per cent and 106 per cent in dry weight and 82 per cent and 55 per cent in plant height respectively. Those values were statistically different from the disease marker (2.77 g and 27.67 cm for weight and height, respectively).

MATERIALS AND METHODS OF FUNGAL ISOLATES

Rhizoctonia solani RBL1, isolated from naturally infected rice plants with typical symptoms of sheath blight in a paddy field of Mazandaran province, Iran, was used in all experiments of this study (obtained from culture collection of Iranian Rice Research Institute). It was purified with the hyphal tip method and maintained on potato dextrose agar (PDA) medium (Merck, Germany). To prove pathogenicity, inoculations were done in glasshouse on Oryza sativa cv. Neda by placing a 5-mm mycelial plug of R. solani between the junction of basal leaf sheath and stem above the water line at maximum tillering stage. R. solani was re-isolated from characteristic lesions of sheath blight. To confirm its anastomosis group (AG) and sub-group, a nuclear rDNA region, containing the ITS1 and 2 as well as the 5.8S gene (accession No. HM211085) was subjected to a BLAST search to find out most similar sequences in NCBI Gen-Bank.

For the isolation of Trichoderma strains, soil samples were collected from rice fields at different geographic locations distributed all over Mazandaran province, which is located on the Southern coast of the Caspian Sea. Soil was taken with a soil auger from 15 cm depth. Samples were air dried for 3–5 days at room temperature. Trichoderma isolates were obtained by the dilution plate method on McFadden and Sutton's RB-S-F Trichoderma selective medium. Sieved soil samples (10 g) were shaken in 90 mL sterile water for 10 minutes. For the isolation of Trichoderma from rice phyllosphere, the leaves and stems were cut into small pieces (1 cm2), transferred into 500 mL Erlenmeyer flask with 100 mL sterile distilled water and placed on a shaker for one hour. A dilution series up to 10-6 was made from the samples. Aliquots (1 mL) were spread onto Petri plates containing selective medium and were then incubated at 25 °C in the dark. Furthermore, Trichoderma isolates were purified directly from observed fungal mass on rice debris. Putative Trichoderma colonies (from soil and foliage samples) were purified on PDA plates by the single spore method and deposited in the Microbiological Collection of the University of Szeged (SzMC).

DNA Extraction and PCR Conditions

For DNA preparation, the mycelium plug of each strain was placed on cellophane disk sterilised by autoclaving in water and placed on the surface of a yeast extract agar medium (5 g of yeast extract, 5 g of dextrose and 20 g of agar per litre) for 2–3 days at room temperature. The fresh mycelium was scraped off and ground with a mortar and pestle in liquid nitrogen. Total DNA was extracted using the Gen-Elute Plant Genomic DNA Miniprep Kit (Sigma, USA) according to the manufacturer's instructions.

A nuclear rDNA region, containing the internal transcribed spacer regions 1 and 2 (ITS1 and 2) and the 5.8S rRNA gene was amplified using the primers ITS1 and ITS4. PCR amplifications were performed as described previously. Amplicons were purified with the Gen-Elute PCR Clean-up Kit (Sigma, USA) and sequenced at Macrogen Inc., South Korea. ITS sequences representing different ITS genotypes were submitted to the NCBI Gen-Bank Database.

Species Identification

A combination of morphological and molecular analyses was used for the identification of Trichoderma isolates. For morphological identification, strains were grown on 2 per cent malt extract agar and on PDA under ambient laboratory conditions of light and temperature (about 21 °C). Microscopic observations and measurements were made from preparations mounted in lactic acid. Individual isolates were identified at the species level using standard mycological keys, species descriptions and TrichOKEY 2.0, an online method for the quick and reliable molecular identification of Hypocrea and Trichoderma at the genus, clade, and species levels based on ITS1 and 2 sequences. TrichOKEY identifies most species unequivocally, although some species share identical or very similar ITS1 and ITS2 sequences and are therefore indistinguishable by the method. An accurate identification of species in the Rufa Clade is sometimes problematic as the species of this clade are currently under revision, and both T. viride and T. koningii are being divided into several new species. In such cases the integration of the barcode method (TrichOKEY) and sequence analysis of other loci with the aid of TrichoBLAST is the preferred strategy of identification.

In vitro Antagonism Experiments

In the case of the dual culture tests, mycelial disks 5 mm in diameter, were cut from the edges of actively growing colonies of the pathogen and Trichoderma isolates, and were separately placed opposite to each other 1.5 cm away from the edge of 9 cm Petri plates containing PDA medium. Plates inoculated with R. solani alone served as control. Each pair was replicated three times and incubated for five days at 26° C in darkness, then scored for degree of antagonism using the rating system of Bell *et. al.* on a scale of 1–5 : Class 1 = Trichoderma completely overgrew the pathogen and covered the entire medium surface, Class 2 = Trichoderma overgrew at least two thirds of the medium surface, Class 3 = Trichoderma and Rhizoctonia each colonised 50 per cent of the medium surface and neither organism appeared to dominate the other, Class 4 = Rhizoctonia colonised at least two-thirds of the medium surface and appeared to withstand encroachment by Trichoderma, and Class 5 = Rhizoctonia completely overgrew the entire medium surface. An isolate of Trichoderma was considered to be antagonistic to the pathogen if the mean score for a given comparison was = 2, but not highly antagonistic if the number was = 3. The plates were maintained for one more week and the number of sclerotia formed was counted and recorded.

Effects of the Trichoderma spp. on germination of R. solani sclerotia were determined according to Mukherjee *et. al.* Briefly, sclerotia of R. solani of uniform size were placed on six days old cultures of the Trichoderma isolates. The cultures were incubated at 26 °C in the darkness for up to 30 days. Untreated sclerotia served as the control. The viability of R. solani sclerotia was evaluated by placing them on the surface of water agar plates for 24 hours at 26 °C and the hyphal growth observed with a stereo-microscope. Sclerotia were considered viable when typical Rhizoctonia-like hyphae grew from the sclerotia.

The effect of volatile metabolites from Trichoderma species against R. solani was tested in the assemblage described by Dennis and Webster. Two bottoms of Petri dishes containing PDA were individually inoculated with a disk of pathogen and Trichoderma, and bottoms were adjusted and sealed by Parafilm. The control sets contain the pathogen alone. The cultures were incubated at 26°C under alternating photoperiod (12 h light/12 h darkness). Growth rates were recorded daily by measuring colony diameter and the inhibition per cent was obtained using the formula : I (per cent)=[(C2–C1)/C2)]×100, where C1 means growth of R. solani in the presence of Trichoderma volatiles, and C2 means growth of the control (R. solani alone).

For the mycoparasitism tests, a small amount of molten water agar was poured and evenly spread over the sterile slide to make a thin agar film. One end of the slide was kept free of the medium to facilitate handling. The mycelial disks of Trichoderma isolates and the pathogen were placed on the slide 1 cm apart from each other. All paired cultures on slides were placed on 2 per cent water agar in Petri dishes and incubated at 28 °C for 5 days.

At the end of the incubation period, regions where the hyphae of Trichoderma isolates met the hyphae of the pathogen were observed under a light microscope for the presence of coiling structures and wall disintegration in the hyphae of Trichoderma and R. solani respectively. The frequencies of coiling or wall disintegration were estimated by counting the coils or areas of disintegration in three different microscopic fields.

Greenhouse Assays

The efficacy of 55 Trichoderma isolates selected based on the *in vitro* screening tests was further evaluated for controlling sheath blight of rice in vivo. The selected isolates belonged to three species *i. e.* T. harzianum, T. virens and T. atroviride. Twenty-days-old rice seedlings (cv. Neda, high yielding and susceptible to sheath blight), were transplanted to plastic pots of 20 cm diameter with 4.5 kg of rice field soil (three seedlings per pot), arranged 25 cm apart on glasshouse benches. They were fertilized by soil application of NPK (100 : 60 : 40 kg/ha) and watered regularly to keep them always submerged. Plants were kept at 26 to 34°C and 85 to 90 per cent relative humidity. The pots were arranged in completely randomised design with three replications for each treatment. The rice plants were inoculated with R. solani as described above (five tillers per hill). Spores of Trichoderma isolates were harvested from 10-days-old culture grown on PDA by adding sterile water to the plates and scraping the culture with a scalpel. These suspensions were filtered through a double layer of cheesecloth to separate large mycelial fragments from conidia.

The concentration of spores was measured with a Neubauer Improved haemocytometer (Precicolour, HBG, Germany) and adjusted to 107 conidia per mL. Tween 20 was added at 0.1 per cent for better adhering of spores to plant surfaces. Rice plants were evenly sprayed with spore suspensions of Trichoderma (50 mL per replicate) using a hand sprayer at two stages : 24 h after inoculation of pathogen and seven days after first spray. Plants inoculated with R. solani alone served as control. Propiconazole (TILT®, 250 EC, Syngenta Crop Protection Inc.), a commonly used fungicide for controlling sheath blight in Mazandaran, was used for comparison.

Disease severity was recorded 30 days after inoculation of the pathogen as Relative Lesion Height (RLH) following the International Rice Research Institute's standard evaluation system for rice (IRRI, 1996). The relative lesion height in each tiller was calculated by using the formula : RLH (per cent) = 100 × [Highest point a lesion is seen (cm)/ Plant height (cm)]. Beside RLH, which was the main parameter recorded as indicative of disease intensity, the number of infected tillers per replicate and the number of sclerotia produced on each rice plant were also counted after 30 days.

Statistical Analysis

All experiments were conducted as completely randomised designs with three replications. Analysis of variance (ANOVA) was performed for each experiment. If significant differences between treatments were detected, means were separated by Fisher's protected least significant differences (LSD) test at P=0.05. Arcsine and square root transformations were used for percentage values and count data, respectively to correct for uneven distribution or homogeneity of variances prior to analyses. All data transformed for analysis are presented as non-transformed values. All analyses were done using the SAS statistics software programme.

Species Identification

Pathogenicity test in glasshouse indicated that R. solani RBL1 was a virulent pathogen on rice. The BLAST search results revealed that this isolate showed highest similarity to six R. solani AG1-IA isolates (Gen-Bank accession numbers: AB000017, AB000016, AB000010, AF354097, AY154301, FJ492099) (E-value = 0.0 and max. identity = 99 per cent).

A total of 202 Trichoderma isolates was obtained as described in the subsection Fungal Isolates of Materials and Methods, and the recovered species comprised Trichoderma harzianum Rifai, T. virens (Miller, Giddens and Foster) von Arx, T. atroviride P. Karst, T. hamatum (Bonord.) Bainier, T. brevicompactum Kraus, Kubicek and Gams and T. asperellum Samuels, Lieckf. and Nirenberg. One hundred and sixteen isolates belonged to the T. harzianum species complex, 70 were identified as T. virens and 11 as T. atroviride. Further, three isolates of T. hamatum, one of T. asperellum and one of T. brevicompactum were obtained. T. harzianum proved to be the dominant Trichoderma species in rice phyllosphere while T. virens was more abundant in soil than in phyllosphere. None of the representatives of the four remaining species were isolated from phyllosphere. In addition, all T. atroviride isolates, except for one, is originating from the eastern part of the Mazandaran province. The number of different species isolated was higher in soil than in the phyllosphere; in fact, from the latter only T. harzianum (31 isolates) and T. virens (three isolates) could be recovered. The population of Trichoderma spp. varied between 0.5×102 to 1×103 cfu/g of dry soil in paddy rice fields.

Dual Culture Tests Between R. Solani and Trichoderma Strains

Antagonism of 78 strains of Trichoderma against R. solani was tested by the dual culture method. None of Trichoderma isolates caused inhibition zone in the colony of R. solani, but most of them were able to overgrow the mycelium of the pathogen and abundantly sporulate on its surface. 83.3 per cent of the examined Trichoderma isolates were antagonistic to R. solani *in vitro*. Sixty two and 37 per cent of the antagonists belonged to T. harzianum and T. virens, respectively. All T. harzianum isolates had a rating of 1 or 2, thus, according to this rating system, all of them proved to be antagonistic.

Sclerotial development of the pathogen was arrested with many Trichoderma isolates and they significantly reduced the number of sclerotia formed by the pathogen. The maximum inhibition occurred when T. harzianum AS15-1 and T. harzianum AS15-4 was paired with R. solani. These two isolates were found to be the most effective, because no sclerotia were produced in culture. There was no correlation between antagonism in dual culture and the ability of isolates in inhibition of sclerotia production *in vitro*. For example, isolate T. virens AD1-5 which belonged to class 3 (non-antagonist), strongly inhibited the sclerotial development of R. solani.

Effect of Trichoderma Strains on the Viability of R. Solani Sclerotia

Seventeen per cent of the Trichoderma isolates displayed complete inhibition of sclerotial germination of R. solani after 30 days. In the case of 23 per cent of isolates, although they did not suppress the germination, they slowed down the growth rate of colonies derived from the sclerotia of the pathogen (reduction ranged from 5.5 to 54 per cent; data not shown).

Surprisingly, 72 per cent of the T. harzianum isolates completely suppressed the germination of sclerotia. Also, 80 per cent of the isolates belonging to this species decreased the colony development from sclerotia.

Effect of Trichoderma Volatile Metabolites on R. Solani

The growth rate of R. solani was reduced significantly by volatile metabolites of Trichoderma isolates (P = 0.05). Results showed that isolate T. harzianum AD6 was the most effective with the highest inhibitory rate of 65.18 per cent followed by T. harzianum AS12-2 and T. harzianum AS3-5 with the rates of 56.66 per cent and 55.55 per cent respectively. Results showed that all T. harzianum isolates demonstrated good level of mycelial growth inhibition of the pathogen. Inversely, ineffective isolates which constituted a statistical group together with the control, belonged to T. virens.

Mycoparasitism Test

Microscopic examination of interaction region in slide cultures clearly showed that at an early state of parasitism (two days after inoculation), 93 per cent of the Trichoderma isolates established close contact with the pathogen by coiling around its mycelia. The coils were usually very dense and appeared to tightly encircle the hyphae of the pathogen. However at this stage of the interaction, integrity of the host cell surface was well preserved. When Trichoderma reached the host hyphae, it usually produced appressorium-like structures or hook-shaped contact branches. After the contact with the host was established, several types of interactions occurred. In one of these, Trichoderma frequently grew parallel to the host and in contact with it. In another common interaction, Trichoderma coiled around the host.

This coiling was either condensed or loose. Pronounced collapse and loss of turgor of R. solani were among the most typical features of alternation observed about 10 days after inoculation. In addition, cell wall breakdown and hyphal disintegration of R. solani hyphae were observed occasionally. Later, R. solani hyphae showed extreme shrinkage and shriveling. T. virens AS16-22 was the most powerful isolate in hyphal disintegrating of the host. There was no correlation between the intensity of coiling and the hyphal disintegration.

Greenhouse Assays

Treatments were significantly different in RLH, number of infected tillers per plants and the number of sclerotia produced by R. solani in the rice phyllosphere.

The disease severity in terms of RLH was significantly reduced by the application of antagonists in comparison to control. Among the isolates, T. harzianum AS12-2 resulted in the lowest RLH and respective disease severity scale, therefore it was the most effective treatment, similarly effective as propiconazole. Furthermore, nine other T. harzianum and three T. virens isolates restricted the disease lesions at lower than 30 per cent of plant height (scale 3). The only T. atroviride isolate was ineffective in this assay. Twenty other Trichoderma isolates constituted the same statistical group with the control inoculated by the pathogen, and were therefore completely ineffective.

The final numbers of infected tillers were recorded one month after inoculation. They ranged from 1 to 13. T. harzianum AS12-2 with the lowest number of infected tillers was more effective than any Trichoderma isolate in confining the horizontal development of R. solani, even more effective than propiconazole, and formed a single statistical group. In the case of T. harzianum AS3-5 and T. virens strains AD1-3 and AS16-22, the number of infected tillers was less than five and these strains established a statistical group together with the fungicide.

The sclerotial number per plant varied from isolate to isolate but the number of sclerotia and disease severity did not show any correlation. However, in the case of treatments where the disease severity scale was 3 or less, the number of sclerotia did not exceed six.

Discussion

Fungi belonging to different species of the genus Trichoderma are very common in soil and also on aerial parts of plants. Based on our results, all examined soil samples contained Trichoderma and the average population was 4.5×102 cfu/g. More than 90 per cent of the Trichoderma isolates obtained from paddy rice fields belonged to T. harzianum and T. virens. Additionally, these were the only species that could be isolated from the above ground parts of rice plant in the fields. Our data indicated that rice field soil may be a favoured habitat for these two Trichoderma species. This finding supports the results of Nagamani and Mew, who reported that T. harzianum is the dominant species in upland rice and a common species in lowland irrigated and rain-fed rice in the Philippines. Mostafa Kamal and Shahjahan isolated 360 Trichoderma strains from rice fields of different geographical locations in Bangladesh and reported that T. harzianum was prevalent in all areas and soil types. Also, Cumagun and Lübeck obtained 42 Trichoderma isolates from rice fields in the Philippines and identified 40 of them as T. harzianum. Interestingly, the most frequent species detected in this study (T. harzianum, T. virens, T. atroviride) were similar to the most abundant Trichoderma spp. found among 69 biocontrol isolates obtained from different geographical locations and culture collections.

Our results demonstrated that Trichoderma strains effectively reduce the growth of R. solani under in vitro conditions. Several mechanisms may be responsible for biocontrol activity of Trichoderma strains. Our study demonstrated the involvement of hyperparasitism and volatile metabolites in the inhibition of R. solani. Cell wall degrading enzymes (CWDEs) such as chitinase, glucanase and proteases are thought to be closely related to the mycoparasitism of Trichoderma strains. Inhibitory volatile substances such as alkyl pyrons were also suggested as the means of biocontrol by some Trichoderma strains.

To be considered a successful BCA, a mycoparasite should be effective against resistant survival structures of plant pathogens. The ability of mycoparasitic fungi to attack sclerotia, thus reducing pathogen inocula in the soil, is one of the potential means of biocontrol. According to the results of the present study, some strains of Trichoderma successfully prevented the production of sclerotia by R. solani *in vitro* and in plant a and also effectively inhibited its germination and growth. This is encouraging for the management of sheath blight, because sclerotia serve as the main primary inoculum that initiates the disease in the next season.

All of the 7 Trichoderma strains most effective in greenhouse assays were antagonistic in dual culture tests according to the system of Bell *et. al.* In contrast, there were no correlations between some *in vitro* biocontrol behaviours of strains and their efficacy in controlling rice sheath blight disease in the greenhouse. For example, T. harzianum AS12-2, the most effective strain in greenhouse assay, could not effectively inhibit the production of sclerotia on Petri plates. In addition, among the 7 biocontrol candidate strains, only T. harzianum AS12-2 suppressed the germination of sclerotia. Moreover, T. harzianum AS12-2 and T. harzianum AS3-5, in spite of controlling the disease, did not coil around the hyphae of the pathogen in the slide culture test.

Application of spore suspension of Trichoderma isolates significantly reduced the disease severity and infected tillers/hill (disease incidence) over control. The results of the present study indicated that foliar spray could be an effective delivery system for controlling the sheath blight of rice. Similar trend was also observed by Tewari and Singh who reported that foliar spray of spore suspension significantly reduces the disease severity and incidence and was found the most effective in controlling sheath blight as compared to soil treatment or seedling root dip.

Furthermore, Rosales and Mew reported that there was significant reduction in lesion development when spore suspensions of Trichoderma spp. were sprayed on rice plants inoculated with R. solani. Verma *et. al.* reported that Trichoderma spp. can be efficiently used as spores, which are more tolerant to adverse environmental conditions during product formulation and field use, in contrast to their mycelial and chlamydospore forms as microbial propagules. This technique for applying Trichoderma in greenhouse proved satisfactory and enabled the fungus to establish itself in the phyllosphere causing a reduction in the impact of R. solani.

Most of the isolates of the genus Trichoderma, which have been found to act as BCAs, have been classified as T. harzianum, leading to the fact that T. harzianum is generally synonymised as a BCA. Moreover, T. harzianum is the active ingredient of many commercial biological control products. The present investigation also showed that most of the Trichoderma strains that were effective both *in vitro* and *in vivo,* belonged to T. harzianum. This is in accordance with the results of Mew and Rosales, Das and Hazarika and Tewari and Singh who reported T. harzianum as an effective BCA in controlling rice sheath blight disease. Moreover, T. harzianum was the most abundant Trichoderma species in rice fields of Mazandaran province and compared to other species, it was able to establish on rice phyllosphere successfully. Therefore we suggest that among the obtained Trichoderma species, T. harzianum is worth for further studying biological control of R. solani and other important causal agents of diseases on rice in northern Iran.

Chapter 22

NUTRITION OF PLANT

The plant nutrients, besides carbon, hydrogen and oxygen, originate from the soil. The soil system is viewed by the soil scientists as a triple-phased system of solid, liquid and a gaseous phases. These phases are physically seperable. The plant nutrients are based in the solid phase and their usual pathway to the plant system is through the surrounding liquid phase, the soil solution and then to the plant root and plant cells.

This pathway may be written in the form of an equation as : M (Solid) $\rightarrow$ M (Solution) $\rightarrow$ N (Plant root) $\rightarrow$ (Plant top) where 'M' is the plant nutrient element in continual movement through the soil-plant system. The operation of the above system is dependent on the solar energy through photosynthesis and metabolic activities.

This is however, an oversimplified statement for gaining a physical concept of the natural phenomenon, but one should bear in mind that there are many physico and physico-chemical processes influencing the reactions in the pathway. The actual transfer in nature takes place through the charged ions, the usual form in which plant-food elements occur in solutions (liquid phase of the system).

Plant roots take up plant-food elements elements from the soil in these ionic forms. The positively charged ions are called 'cations' which include potassium(K^+), Calcium(Ca^{++}), magnesium(Mg^{++}), iron(Fe^{+++}), zinc(Zn^{++}), and so on. The negatively charged ions are called anions and the important plant nutrients taken in this form include nitrogen(NO^-_3), phosphorous($H_2PO^-_4$), sulphur(SO^-_4), Chlorine(Cl), etc.

The process of nutrient uptake by plants refers to the transfer of the nutrient ions across the soil root interfaces into the plant cell. The energy for the process is provided by the metabolic activity of the plant and in its absence no absorption of nutrients take place. Nutrient absorption involves the phenomenon of ion exchange. The root surface, like soil, carries a negative charge and exhibits cation-exchange property. The most efficient absorption of the plant nutrients takes place on the younger tissues of the roots, capable of growth and elongation.

The key-role of fertilizers and their judicious use in crop husbandry is well understood, when one is familiar with the general facts about plant nutrition. It is now known that at least 16 plant-food elements are necessary for the growth of green plants. These plant-nutrients are called essential elements. In the absence of any one of these essential elements, a plant fails to complete its life cycle, though the disorder caused can, however, be corrected by the addition of that element.

These 16 elements are : Carbon(C), hydrogen(H), oxygen(O), nitrogen(N), phosphorous(P), sulphur(S), potassium(K), calsium(Ca), magnesium(Mg), iron(Fe), manganese(Mn), zinc(Zn), copper(Cu), molybdenum(Mb), boron(B) and chlorine(Cl). Green plants obtain carbon from carbon-di-oxide from the air; oxygen and hydrogen from water, whereas the remaining elements are taken from the soil.

Based on their relative amounts, normally found in plants, the plant nutrients are termed as macro-nutrients, if large amounts are involved, and micro-nutrients, if only traces are involved. The micro-nutrients essential for plant growth are iron, manganese, copper, zinc, boron, molybdenum, and chlorine. All other essential elements listed above are macro-nutrients.

In this respect, root-systems are known to vary from crop to crop. Hence their feeding power differs. The extent and the spread of the effective root-system determines the soil volume trapped in the feeding-zone of the crop plant. This is indeed an important information in a given soil-plant system which helps us to choose fertilizers and fertilizer-use practices. The absorption mechanisms of the crop plants are fairly known now. There are three mechanisms in operation in the soil-water-plant systems.

They are :

- The contact exchange and root interception,
- The mass flow or convection, and
- *Diffusion* : In the case of contact exchange and root interception, the exchangeable nutrients ions from the clay-humus colloids migrates directly to the root surface through contact exchange when plant roots come into contact with the soil solids.

Nutrient absorption through this mechanism is, however, insignificant as most of the plant nutrients occur in the soil solutions. Scientists have found that plant roots actually grow to come into contact with only 3 per cent of the soil volume exploited by the root mass, and the nutrient uptake through root interception is even still less.

The second mechanism is mass flow or convection, which is considered to be the important mode of nutrient uptake. This mechanism relates to nutrient mobility with the movement of soil water towards the root surface where absorption through the roots takes place along with water. Some are called mobile nutrients. Others which move only a few millimetres are called immobile nutrients. Nutrient ions such as nitrate, chloride and sulphate, are not absorbed by the soil colloids and are mainly in solution. Such nutrient ions are absorbed by the roots along with soil water.

The nutrient uptake through this mechanism is directly related to the amount of water used by the plants (transpiration). It may, however, be mentioned that the exchangeable nutrient cations and anions other than nitrate, chloride and sulphate, which are absorbed on soil colloids are in equilibrium with the soil solution do not move freely with water when it is absorbed by the plant roots.

These considerations, therefore, bring out that there are large differences in the transport and root absorption of various ion through the mechanism of mass flow. Mass flow is, however, responsible for supplying the root with much of the plant needs for nitrogen, calcium and magnesium, when present in high concentrations in the soil solution, but does not do so in the case of phosphorous or potassium. The nutrient uptake through mass flow is largely dependent on the moisture status of the soil and is highly influenced by the soil physical properties controlling the movement of soil water.

The third mechanism is diffusion. It is an important phenomenon by which ions in the soil medium move from a point of higher concentration to a point of lower concentration. in other words, the mechanism enables the movement of the nutrients ion without the movement of water. The amount of nutrient-ion movement in this case is dependent on the ion-concentration gradient and transport pathways which, in turn, are highly influenced by the content of soil water. This mechanism is predomionant in supplying most of the phosphorous and potassium to plant roots.

It is important to note that the rhizophere volume of soil in the immidiate neighbourhood of the effective plant root receives plant nutrients continously to be delivered to the roots by diffusion. However, when the nutrient concentration builds up far excess of the plant in the reverse direction. These are some of the choice of fertilizers and fertilizer practices for practising scientific agriculture.

The relationship in the soil-plant system stated in the simple equation give in the earlier paragraph reflects the highly dynamic nature of the soil solution. One knows that the roots of the growing plants continuously remove nutrient ions from the soil solutions. At the same time, the breakdown of the soil minerals and the generating of more exchangeable cations, the biological activity and the additions made to the anions, *e.g.* nitrates, continuously change the composition of the soil solution. At a given point of time, therefore, the available plant nutrients in the soil solution may range from a tiny amount to larger quantities. Under favourable conditions, crop plants, in general, require larger amounts of plant nutrients than the quality found in soil solution at any given time.

Hence, the situation of nutrients supply to plants becomes a limiting factor, specially, at the critical stages of plant growth and low crop yeilds result in recognition, therefore, fertilizers application and the use of suitable fertilizers are recommended for higher crop yeilds in productive farming. The knowledge of the specific role of each essential element in the growth of crop plants and their amounts required for efficient crop production is considered necessary in adopting scientific fertilizers use.

Plant Nutrients and Their Functions

The plants require, the following essential nutrients for their normal development :

Carbon	Nitrogen	Calcium
Hydrogen	Phosphorous	Magnesium
Oxygen	Potassium	Sulphur
Iron	Zinc	Chlorine
Manganese	Boron	..
Copper	Molybdenum	..

Carbon is obtained from carbon dioxide of the air; Oxygen from air and water; Hydrogen from water; Nitrogen from air and soil or both, and all other nutrients from the soil. Soil is a the most important source of plant food.

Nitrogen, Phosphorous and Potassium are known as primary plant nutrients; Calcium, Magnesium and Sulphur are secondary nutrients; Iron, Manganes, Copper, Zinc, Boron, Molybdenum and Chlorine as trace elements or micro-nutrients.

The primary nutrients and secondary nutrients elements are known as major elements. This classification is based on their relative abundance, and not their relative importance. the micro-nutrients are required in small quantities, but they are as important as the major elements in plant nutrients.

Air is the primary source of Nitrogen for plant nutrient. Only leguminous crops can directly use this free Nitrogen with the help of symbiotic bacteria of the genus *Rhizobium*. Other plant derive from soil their Nitrogen in the form of Nitrogen and Ammonium.

Nitrogen and Ammonium are produced in the soil by action of micro-organisms on the soil organic matter. Non symbiotic micro-organisms can fix free Nitrogen of the air and make it available to plant in Ammonium and Nitrates forms.

Nitrogen : Nitrogen encourages the vegetative development of plants by importing a healthy green colour to the leaves. It also controls, to some extent the efficient utilisation of phosphorous and Potassium. Its dependency retards growth and root development, turns the foilage yellowish or pale green, histens maturity, causes the shrivelling of grains and lowers crop yield.

The older leaves are affected first. An excess of Nitrogen produces leathery (sometimes crinkled), dark-green leaves and succulent growth. It also delays the maturation of plants, impairs the quality of crops like barley, potato, tobacco, sugarcane, and fruits; increases susceptibility to diseases and causes 'lodging' of cereal crops by inducing an undue lengthoning of the stem internodes.

Phosphorous : *Phosphorous* influences the vigour of plants and improves the quality of crops. It encourages the formation of new cells, promotes root growth (particularly the development of fibrous roots), and hastens leaf development through emergence of ears, the formation of grains, and the maturation of crops. It also increases resistence to diseases and strengthens the stems of cereal plants, thus reducing their tendency to lodge.

It offsets the harmful effects of excess nitrogen in the plant. When applied to leguminous crops, it hastens and encourages the development of nitrogen-fixing nodule bacteria. If phosphorous is deficient in the soil, plants fail to make a quick start, do not develop a satisfactory root-system, remain stunted and sometimes develop a tendency to show a reddish or purplish discolouration of the stem and foilage owing to an abnormal increase in the sugar content and the formation of anthoscyanin.

However, the deficiency of this element is not easily recognised as that of nitrogen. It has also been observed that cattle feeding on the produce of deficient soils become dwarfed, develop stiff joints and lose the velvetty feel of the skin. Such animals show an abnormal craving for eating bones and even soil itself.

Potassium : *Potassium* enhances the ability of plants to resist diseases, insect attacks, and cold and other adverse conditions. It plays an essential part in the formation of starch and in the production and translocation of sugars, and is thus of special value to carbohydrate-rich crops, *e.g.* sugarcane, potato and sugar-beet. The increased production of starch and sugar in legumes fertilized with potash benefits the symbiotic bacteria and thus enhances the fixation of nitrogen.

It also improves the quality of tobacco, citrus etc. With an adequate supply of potash, cereals produce plump grains and strong straws. But an excess of element tends to delay maturity, though, not to be the same extent as nitrogen.

Plants can make up and store potassium in much larger for correcting zinc deficiency. The symptoms of zinc deficiency appear generally in younger leaves, starting with intervienal chlorosis leading to a reduction in shoot growth and the shortening of internodes. Mottle leaf, little leaf, etc., in the case of trees are symptoms of zinc difficiency.

The buds of several defficient maise plants become white; in citrous interveinal chlorosis and mottled leaf occur. In calcareous soils and in soils with very high phosphorus content, zinc defficiency is commonly expected to occur. The principal function of zinc in plants is as a metal activator of enzymes. In highly weathered coarse textured soils zinc deficiency appears under an intensive cropping programme.

The availability of zinc is least between pH 5.5 and 7, but its availability increases at a lower pH. At a higher pH above 7, zinc availability becomes a complex problem, as the positively charged zinc ion gets converted into a negatively charged zincate complex whose availability tends to be reduced in alkaline soils.

When the calcium ion is predominent, the highly insoluble calcium zincate is formed and zinc availability gets seriously limited. The application of soluble zinc salts or zinc chelates to the soil is generally recommended to correct its defficiency. Foliar sprays are advocated, especially for orchard trees for amending zinc defficiency. About 5 to 50 kg of zinc sulphate per hectare is used for such purposes.

The symptoms of *boron* defficiency vary with the kind and age of the plant, the conditions of growth and the severity of the deffiency. Each crop produces its characteristic growth abnormalities associated with boron defficiency, such as yellows and rosetting in lucerne, snakehead in wallnuts, die-back and corking of fruits in apple, corking and pitting of fruits in tomatoes, hollow stem and the bronsing of curd in cauliflower, the brown-heart diseases in table-beets, turnips, etc.

Molybdenum deficiency produces whip-tail in cauliflower, broccoli and other *Brassica* crops. The deficiency of this element reduces the activity of the symbiotic and non-symbiotic nitrogen-fixing micro-organisms.

It was in 1954 that *chlorine* was proved to be an essential micronutrient. Its defficiency under field conditions has not been reported so far. In water-culture solutions, the leaves of chlorosis, necrosis and an unusual bronze discolouration on tomatoes.

Sodium is not an essential element for plant growth. But some crops, such as beet, celery, cabbage, kale, knol-khol, radish, rape and turnip, benefit greatly by application of soluble sodium salts, specially if the soil is deficient in potassium. Sodium is also of direct benefit to plants indigneous to the sea-shore or to irrigated arid regions. Salts of this element are said to release more of potassium from the exchange complex and to help to maintain phosphorus in a more available form. They also serve as a partial substitute for potassium in the case of potatoes and cotton.

Maintenance of Soil Fertility

No two soils are alike either in respect of their nature or in respect of quantities of plant nutrients they contain. Under a given situation, the system of farming, soil management and manuring practices, etc., influence the productivity of soils and crop yields obtained from them.

It is estimated that the different agricultural crops in India remove about 4.27 million tonnes of nitrogen, 2.13 million tonnes of phosphoric acid, 7.42 million tonnes of potash and 4.88 million tonnes of lime per year. The production of larger yields through improved varities of crops and intensive cultivation will increase the depletion of nutrients still further. But erosion and leaching cause additional losses.

The present production of synthetic nitrogenous fertilizers in the country reached 1.5 million tonnes of nitrogen, and the bulky organic manures might supply another 1.5 million tonnes of total nitrogen. The amounts of phosphorous, potash, etc., added to the soil are very small. It is thus obvious that the current huge drain on nutrient supplies will continue to impoverish the soils unless these supplies are replenished by natural or by artificial means, the principal methods of supplementing natural recuperation and for improving the productive capacity of the soils are :

- To add organic matter to the soil, so that through decay, it may furnish a more or less continuous supply of nuttrients for crops,

To restore or increase the amount of defficient nutrients by the application of fertilizers.

The urgent need of the constantly expanding agriculture production to meet the requirements of the continually increasing human and cattle populations in India makes the supply of additional plant nutrients through fertilizers and organic manures, a problem of supreme importance.

Types of Manures and Fertilizers

Indian soils are usually very poor in organin matter as well as in nitrogen. Pohsphate deficiency is less wide-spread and potash deficiency generally occurs in compact areas. In acid soils the addition of lome steps up production. Materials which are commonly used to maintain and improve soil fertility.

- *Manures* : These are relatively bulky materials, such as animal or green manures, which are added mainly to improve the physical condition of the soil, to replenish and keep up its humus status, to maintain the optimum conditions for the activities of soil micro-organisms and make good a small part of the plant nutrients removed by crops or otherwise lost through leaching and soil erosion.
- They, thus, supply practically all the elements of fertility which crops require, though not in adequate proportions. The plant-food elements contained in a manure are released in an available form after it is applied to the soil and is decomposed by soil micro-organisms. Similarly, the green manures add not only substantial amounts of organic matter but also nitrogen.
- *Fertilizers* : Fertilizers are inorganic materials of a concentrated nature; they are applied mainly to increase the supply of one or more of the essential nutrients, *e.g.* nitrogen, phosphorous and potash. Fertilizers contain these elements in the form of soluble of readily available chemical compounds. This distinction is, however, not very rigid. In common parlance the fertilizers are sometimes called 'chemical','artificial' or 'inorganic' manures.
- *Concentrated Organic Manures* : Some of the concentrated materials, such as oil-cakes, bone-meal, urine and blood are of organic origin. The use of manures and fertilizers is complementary and not as a substitute for each other.
- *Bulky Organic Manures* : The properties and role of organic matter and humus in the soil have been explained already.

Farmyard Manure

Good-quality farmyard manure is perhaps the most valuable organic matter applied to a soil. It is the most commonly used organic manure in India. It consists of a mixture of cattle dung, the bedding, used in the stable and of any ramnants of straw and plant stalks fed to cattle. Though its crop-increasing value has been recognised from time immemorial, more than 50 per cent of the cattle dung produced in the country today is burnt as fuel and is thus lost to agriculture. Not only this tremendous waste, but also the tradition method of preparing and storing the farmyard manure is generally faulty.

The cattle-dung, together with stable-waste and house sweeping, is first collected in the open backyard, and when a cartload has been collected, it is removed to another heap or to an uncovered pit in a common plot outside the village. The loose heaps lie exposed to the sun, with the result that the raw organic matter dries up quickly and does not rot properly.

Very often, a part of the dry dung is blown off by wind or washed away by rain. Cattle urine is either not conserved or is stored in a defective manner. American studies on the distribution of soil derived elements between urine and faeces of dry cows have shown that 95 per cent of Potassium, 63 per cent of nitrogen and 50 per cent of sulphur are contained in the urine.

The wastage of nitrogen-rich urine, the loss of nitrogen (in the form of ammonia) due to the fermentation of exposed cattle dung, and the washing away of soluble mineral elements be leaching reduce its manurial value in India to a great extent. Its average content of plant nutrients under Indian and European conditions is shown below for comparison :

Percentage Content

	N	P_2O_5	K_2O
India	0.3	0.15	0.3
European countries	1.0	0.30	1.0

About half of this nitrogen, one-sixth of phosphorous and more than half of potash are readily soluble and subject to dissipation. However, the loss of nitrogen and minerals elements caused by careless handeling can be reduced greatly by using absorbent bedding for cattle, storing dung in stone or brick-line pits, mixing large quantioties of straw and other vegitable matter with cattle dung, and keeping the heap compact and moist. Thus, if urine is properly conserved, the loss of soluble mineral elements through seepage is prevented, bacterial decomposition of raw organic matter is encouraged, plant nutrients are made soluble, and nitrogen losses are minimised.

Table. The Relative Absorbent Capacity of Different Materials used as Bedding for Cattle.

Material	Quantity of water (in kg) retained by one kg of the following materials after 24 hours of soaking
Wheat straw	2.20
Peat straw	2.80
Dry leaves	2.00
Peat	6.00
Sawdust	4.35
Soil	0.50
Sand	0.25

If urine is not conserved in the bedding used for cattle it must be collected in covered *pucca* cistern and then, added to the dung in the manure pit. Nitrogen in the urine is mainly in the form of urea, which readily changes into the highly volatile ammonium carbonate through bacterial action, and quickly loses ammonia thereafter by evapouration. This loss can be reduced to a great deal if the manure and the urine-soaked absorptive itter for bedding are kept compacted in a pit.

The pit may be 1 m in depth, 1.3 to 1.5 m in width and 4.5 to 6 m in length, depending upon the No. of cattle on a farm. The filling of the pit should be 'sectional' and when each section of three or 1.3 m in length is filled to about 45 cm above the ground level, it should be clustered with 2.5 cm layer of a mixture of mud and dung in equal proportions. Before plastering, 4 to 5 buckets of water should be added to the manure in the pit. Plastering conserves moisture and nitrogen and also prevents housefly nuisance. The manure becomes ready for use in about 4 to 5 months after plastering.

The quality of manure is also improved by the concentrated feeds given to cattle. Cotton-seed, cotton-seed cake, linseed-meal, wheat bran, grain husk, groundnut cake, gram, horse-gram, etc., are rich in nitrogen, phosphorous, potassium, magnesium and sulphur.

It has been found that in the case of adult working-cattle about 80 per cent of nitrogen and the other mineral elements contained in the feed is recovered in urine, faeces and other animal by-products. Accordingly, manure from cattle fed on cereal straws and grass hay is much less valuable than that from animals fed on legume hays, grains and concentrates.

In foreign countries, considerable attention has been given to the use of preservatives on manure. Calcium sulphate or gypsum and superphosphate have proved most promising in preventing the escape of ammomnia. Gypsum has been found specially effective as an ammonia-absorbing agent. Superphosphate, besides absorbing ammonia, supplies additional phosphorous and, thus improves the crop producing capacity of the manure.

Partially rotten farmyard manure should generally be applied to the soil about three to four weeks before the sowing of a crop. In case there is sufficient moisture in the soil, there sill be enough time for its decomposition and for improving the soil structure. Its application too long before sowing a crop will either cause a drying up of the rotted manure or too quick decomppostion, depending on the incidence of rains. But in each case, there will be a serious loss of ammonia and nitrogen. If the manure is already well rotted, it is advisable to apply it just before sowing to a crop.

This procedure is perticularly essential in the case of light soils. In any case, after the manure is carted to the field, it should be evenly spread and worked into the soil soon to avoid the loss of nitrogen. The existing practice of leaving the manure in small heaps scattered in the field for several days, before spreading it in the field and incorporating it into the soil, results in the serious deterioration of its quality, particularly if strong winds blow.

In vagitable and fruit cultivation, the application of well-rotten manure, in conjunction with fertilizers to young plants individually, has been founnd to give the best results. In Egypt, even the cotton crop, which is invariably grown on ridges, is given a top-dressing a handful of the rotten manure being applied to the soil at the base of each plant before an irrigation, and is worked into the soil with a hand hoe.

The practice of penning cattle, sheep and goats in the fields in summer is common in some parts of the country. Folding 7,000 sheep for one night is said to add the equivalent of 149.3 quintals (14.93 tonnes) of cattle dung. The fresh dung left in the field in such cases rapidly dries up. This drying checks ammonification and loss of nitrogen. With the first fall of rain, the dung is worked into the soil. It, therefore, does not lose much of its fertilizing value. Further more, its beneficia effects on the physical condition of the soil are undeniable. However, sheep-folding is said to make the land more weedy.

There must be adequate moisture in the soil for the proper decomposition of organic matter. Farmyard manure can, therefore, be applied to all crops grown in the rainy season or grown under irrigation. The quantity of manure to be applied to unirrigated crops varies from 1.5 to 2 cartloads per hectare in areas of heavier rainfall.

If sufficient farmy manure is not available, it may be applied at the usual rate to a part of the land, say, to one-third or one-forth of the area, in rotation every year, so that all parts of the field receive the manure regularly once in three or four years. For irrigated field crops, the rate varies from 10 to 20 cartloads. Sugarcane, maise and garden crops, such as potatoes, turmeric, ginger, vegetables and fruits receive still higher doses, amounting sometimes to 15 to 25 cartloads. A cartload of manure, measuring 9 cubic metres, weighs about half a tonne.

It must be stressed that the value of farmyard manure in soil improvement is due to its content of principal nutritive elements and its ability to :

- Improve the soil tilth and aeration,
- Increase the water-holding capacity of the soil,

Stimulate the activity of micro-organisms that make the plant-food elements in the soil readily available to crops.

The supply of organic matter, which is later converted into humus is a property of farmyard manure.

One tonne of cattle dung can supply only 2.95 kg of nitrogen, 1,59 kg of phsophoric acid and 2,95 kg of potash.

The use of farmyard manure alone causes an imbalance in nutirtion owing to its relatively low content of phosphoric acid. Therefore, to keep the soils well supplied with all the essential elements of plant food in a readily available form, and also to keep them in good 'heart', it is advisable to use the bulky organic manures in conjuction with superphosphate and such other artificial fertilizers as contain the particular plant food or foods in which a soil may be deficient or which the crop to be grown may specially require.

Chapter 23

ORGANIC FORM OF SOIL

NITROGEN

Over 90 per cent of the nitrogen N in the surface layer of most soils occurs in organic forms, with most of the remainder being present as NH_4^- which is held within the lattice structures of clayminerals. The surface layer of most cultivated soils contains between 0.06 and 0.3 per cent N. Peat soils have high N contents to 3.5 per cent. Plant remains and other debris contribute nitrogen N in the form of :

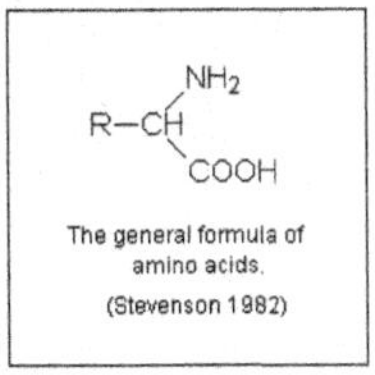

Fig. Amino Acids.

Amino acids exist in soil in several different forms, like :

- As free amino acids
  - In the soil solution
  - In soil micropores
- As amino acids, peptides or proteins bound to clay minerals
  - On external surfaces
  - On internal surfaces
- As amino acids, peptides or proteins bound to humic colloids
  - H-bonding and van der Waals' forces
  - In covalent linkage as quinoid-amino acid complexes
- As mucoproteins
- As a muramic acid.

Amino acids, being readily decomposed by microor-ganisms, have only an-ephemeral existence in soil. Thus the amounts present in the soil solutionat any one time represent a balance between synthesis and destruction by micro-organisms.

The free amino acids content of the soil is strongly influenced by weathercon-ditions, moisture status of the soil, type of plant and stage of growth, additions of organic residues, and cultural conditions.

Amino Sugars

Amino sugars occur as structal components of a broad group of substances, the mucopolysaccharides and they have been found in combination with muco-peptides and mucoproteins. Some of the amino sugar material in soil may exist in the form of an alkali-insoluble polysaccharide referred to as chitin. Generally the amino sugars in soil are of microbial origin. From 5 to 10 per cent of the N in the surface layer of most soils can be accounted for in N-containing carbohydrates or amino sugars.

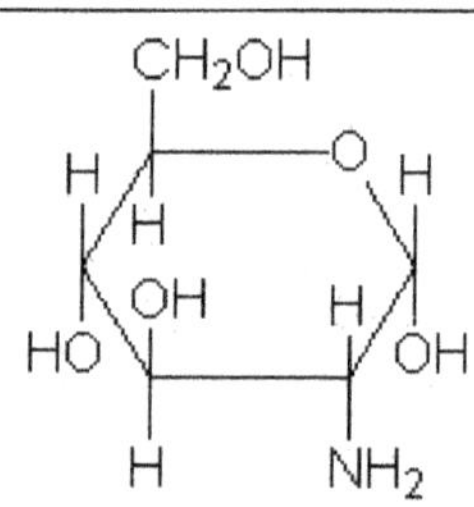

D - Glucosamine
(Stevenson 1982)

Nucleic Acids

Nucleic acids, which occur in the cells of all living organisms, consist of individual mononucleotide units (base-sugar-phosphate) joined by a phosphori-cacid ester linkage through the sugar.Two types : ribonucleic acid (RNA) andde-oxyribonucleic acid (DNA).They have pentose sugar (ribose or deoxyribose), the purine : adenine, guanine and the pyrimidine : cytosine, thymine. RNA contains also the uracil.

The N in purine and pyrimidine bases is usually considered to account for-less than 1 per cent of the total soil N. Small amounts of N are extrcted from soil in the form of glycerophosphatides, amines, vitamins, pesticide and pesticide degradation products.

Nitrogen Transformation

A key feature of the internal cycle is the biological turnover of N between-mineral and organic forms through the opposing processes of mineralisation and immobilisation. The latter leads to incorporation of N into microbial tissues.

Whereas much of this newly immobilised N is recycled through mineralisation, some is converted to stable humus forms. The overall reaction leading to incorporation of inorganic forms of N intostable humus forms is depicted on the picture.

Thus the decay of plant and animal residues by micro-organisms results in theformation of mineral forms of N (NH_4^+ and NO_3^-) and assimilation of part of the C into microbial tissue (reaction A). Part of the native humus undergoes a similar fate (reaction B). Subsequent turnover through mineralisation-immobilisation leads to incorporation of N into stable humus forms (reaction C). Stabilisation of N may also occur through the reaction of partial decay products of lignin with nitrogenous constituents (reaction D). Except under unusual circumstances, both mineralisation and immobilisationalways function in soil, but in opposite direction.

Chemical Reaction of Ammonia and Nitrite with Organic Matter

The fate of mineral forms of N in soil is determined to some extent bynon-biological reactions involving NH_4^+, NH_3 and NO_2^- as depicted in fig.

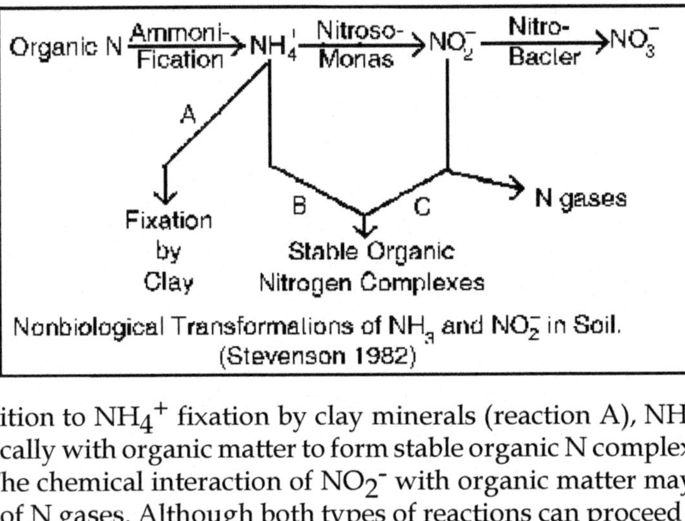

Nonbiological Transformations of NH_3 and NO_2^- in Soil.
(Stevenson 1982)

In addition to NH_4^+ fixation by clay minerals (reaction A), NH_3 and NO_2^- react chemically with organic matter to form stable organic N complexes (reaction B and C). The chemical interaction of NO_2^- with organic matter may lead to the generation of N gases. Although both types of reactions can proceed over a wide pH range, fixation of NH_3^+ is favoured by a high pH (>7.0). In contrast, NO_2^- -organic matter interactions occur most readily under highly acidic conditions (pH of 5.0 to 5.5 or below).

Stability of Soil Organic Nitrogen

- Proteinaceous constituents are stabilised through their reaction with other organic constituents, such as lignins, tanins, quinons.
- Biologically resistant complexes are formed in soil by chemical reaction involving NH_3^+ or NO_2^- with lignins or humic substances.
- Adsorption of organic N compounds by clay minerals (paricular y montmorillinitic types) protects the molecule from decomposition.

- Complexes formed between organic N compounds and polyvalent cations, such as Fe, are biologically stable.
- Some of the organic N occurs in small pores or voids and is physicallyi-naccessible to micro-organisms.

C/N Ratio

For surface soils, and for the top layer of lake and marine sediments, the ratio generally falls within well-defined limits, usually from about 10 to 12. In most soils, the C/N ratio decreases with increasing depth, often attaining values less than 5.0. Native humus would be expected to have a lower C/N ratio than most undecayed plant residues for following reasons. The decay of organic residues by soil organisms leads to incorporation of part of the C into microbial tissue with the remainder being liberated as CO_2.

As a general rule, about one-third of the applied C in fresh residues will remain in the soil after the first few months of decomposition. The decay process is accompanied by conversion of organic form of N to NH_3 and NO_3^- and soil micro-organisms utilise partof this N for synthesis of new cells.

The gradual transformation of plant raw material into stable organic matter (humus) leads to the establishment of reasonably consistent relationship between C and N. Other factors which may be involved in narrowing of the C/N ratio include chemical fixation of NH_3 or amines by lignin like substances.

The C/N ratio of virigin soils formed under grass vegetation is normally lower than for soils formed under forest vegetation, and for the latter, the C/N ratio of the humus layers is usually higher than for the mineral soil proper.

Also the C/N ratio of a well-decomposed muck soil is lower than for a fibrous peat. As a general rule it can be said that conditions which encourage decomposition of organic matter result in narrowing of the C/N ratio. The ratio nearly always narrows sharply with depth in the profile; for certain subsurface soils C/N ratios lower than 5 are not uncommon.

ORGANIC PHOSPHORUS

Phosphorus rank in importance with N and K as major plant nutrients.

Phosphorus compounds in soil can be placed into the following three classes :

1. organic compound of the soil humus,
2. inorganic compounds in which the P is combined with Ca, Mg, Fe, Al and with clay minerals,
3. organic and inorganic P compounds associated with the cells of living matter. Micro-organisms are involved in transformations of phosphorus between organic and mineral forms.

From 15 to 80 per cent of the phosphorus in soils occurs in organic forms, the exact amount being dependent upon the nature of the soil and its composition.

The higher percentages are typical of peats and uncultivated forest soils. From the standpoint of plant nutrion, phosphorus is adsorbed by plants largely as the negatively charged primary and secondary orthophosphate ions ($H_2PO_4^-$ and HPO_4^{2-}) which are present in the soil solution. Small quantities of soluble organic P compounds are also present in water extracts of soil.

The Phosphorus Cycle

In a broad sense, the phosphorus cycle in soil involves the uptake of phosphorus by plants and its return ti the soil in plant and animal residues.

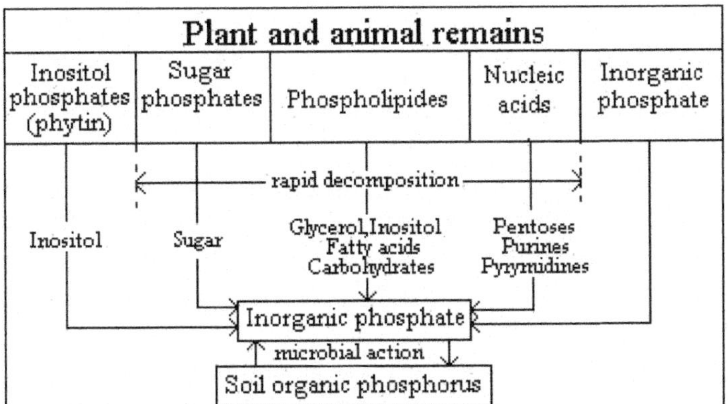

Phosphorus transformations in soil. According to W. Flaig (1966).

As can be seen from picture three general types of compounds make up the bulk of the organic phosphorus in plants, namely : phytin, phospholipids, and nucleic acids. Approximate recoveries of organic phosphorus in these forms are as follows :

Inositol phosphates	2-50 per cent
Phospholipids	1-5 per cent
Nucleic acids	0.2-0.5 per cent
Phosphoproteins	trace
Metabolic phosphates	trace

When crop residues are returned to the soil, net immobilisation of P will occur when the C/organic- P ratio is 300 or more; net mineralisation will result when the ratio is 200 or less.

FUNCTION OF ORGANIC MATTER IN SOIL

Organic matter contributes to plant growth through its effect on the physical, chemical, and biological properties of the soil.

It has a :

- *Nutritional* function in that it serves as a source of N, P for plant growth

- *Biological* function in that it profoundly affects the activities of microflora and microfaunal organisms
- *Physical* and *physico-chemical* function in that it promotes good soil structure, thereby improving tilth, aeration and retention of moisture and increasing buffering and exchange capacity of soils.

Humus also plays an indirect role in soil through its effect on the uptake of micro-nutrients by plants and the performance of herbicides and other agricultural chemicals.

It should be emphasised that the importance of any given factor will vary from one soil to another and will depend upon such environmental conditions as climate and crpping history.

Availability of Nutrients for Plant Growth

Organic matter has both a direct and indirect effect on the availability of nutrients for plant growth. In addition to serving as a source of N, P, S through its mineralisation by soil micro-organisms, organic matter influences the supply of nutrients from other sources (for example, organic matter is required as an energy source for N-fixing bacteria).

A factor that needs to be taken into consideration in evaluating humus as a source of nutrient is the cropping history. When soils are first placed under cultivation, the humus content generally declines over a period of 10 to 30 years until a new equilibrum level is attained. At equilibrium, any nutrients liberated by microbial activity must be compensated for by incorporation of equal amounts into newly formed humus.

Effect on Soil Physical Condition, Soil Erosion and Soil Buffering and Exchange Capacity

Humus has a profound effect on the structure of many soils. The deterioration of structure that accompanies intensive tillage is usually less severe in soils adequately supplied with humus. When humus is lost, soils tend to become hard, compact and cloddy.

Aeration, water-holding capacity and permeability are all favourably affected by humus. The frequent addition of easily decomposable organic residues leads to the synthesis of complex organic compounds that bind soil particles into structural units called aggregates. These aggregates help to maintain a loose, open, granular condition. Water is the better able to infiltrate and percolate downward through the soil.The roots of plants need a continual supply of O_2 in order to respire and grow. Large pores permit better exchange of gases between soil and atmosphere.

Humus usually increases the ability of the soil to resist erosion. First, it enables the soil to hold more water. Even more important is its effect in promoting soil granulation and thus maintaining large pores through which water can enter and percolate downward. From 20 to 70 per cent of the exchange capacity of many soils is caused by colloidal humic substances. Total acidities of isolated fractions of humus range from 300 to 1400 meq/100g. As far as buffer action is concerned, humus exhibits buffering over a wide pH range.

Effect on Soil Biological Condition

Organic matter serves as a source of energy for both macro- and micro-faunal organisms. Numbers of bacteria, actinomycetes and fungi in the soil are related in a general way to humus content. Earthworms and other faunal organisms are strongly affected by the quantity of plant residue material returned to the soil. Organic substances in soil can have a direct physiological effect on plant growth. Some compounds, such as certain phenolic acids, have phytotoxic properties; others, such as the auxins, enhance plant growth. It is widely known that many of the factors influencing the incidense of pathogenic organisms in soil are directly or indirectly influenced by organic matter. For example, a plentiful supply of organic matter may favour the growth of saprophytic organisms relative to parasitic ones and thereby reduce populations of the latter. Biologically active compounds in soil, such as antibiotics and certain phenolic acids, may enhance the ability of certain plants to resist attack by pathogens.

TYPES OF HUMUS IN SOILS

Humus occurs in soils in many types, differentiates in regard to morphology and fractional composition. A type of humus is it a morphological form of naturals accumulation of humic substances in profile or on the surface of soil, conditioned by general direction of soil-forming process and humification of organic matter.

A types of humus in terrestrial enviroment are following :

- Mor
- Moder
- Mull.

Mor is a type of humus, which occur largely in coniferous forest soils and the moorlands soils. This humus arise under conditions of low-biological activity in soil. The mineralisation of organic matter proceed slowly and create layers, which maintain a structure of vegatable material. Acidophilic fungi and low activeinvertebrates participates in transformations of plant residues.

Under these circumstances forms a litter of large thickness. C/N ratio of mor humus is always more than 20, or even 30-40, whereas pH is acid. Moder is a transitional form of humus between mull and moder, characteristic for sod-podzolic soils, loesses and mountain grassland soils.

The organic horizons with moder humus consist of low-thicknessed litter (2-3 cm), which gradually, without bounds, pass on to humus-accumulative horizons. Moder is a type of medium humified humus. Acidophilic fungi and arthropodan participates in transformations of plant residues. C/N ratio equal 15-25. Produced mineral-organic complexes are labile and weakly bounded with mineral portion of soil.

Mull is a type of humus characteristic for chestnut soils, phaeozems, rendzinas and others soils. This type of humus arise under grass vegetation. Mull is a well humified organic matter, which is produce in very biologically active habitat.

This type of humus is characterised by neutral pH, C/N ratio nearing to 10 and ability to creation stable mineral organic complexes. Mull is a type of humus which occurs in soils under cultivation. According to Kononova, the types of humus are divide as follows : First type of humus is characteristic for podzolic soils, grey brown soils and lateric soils under forest communities. In this humus predominate humic acids, thus humic acid/fulvic acid ratio is below 1. Humic acid indicate small extent of aromatic rings condensation and they are approximate to fulvic acids. Considerable hydrophilic properties of humic acids favour to creation of chelates with polyvalent cations and ability to displacement deep into profile of soil. Considerable mobility of this humus favour process of podsolisation. Second type of humus is characteristic for phaeozems, rendzinas, black earths and brown soils. Humic acid/fulvic acid ratio is upper than 1, Extent of aromatic rings condensation is high in humic acids, which cause their hydrophobic properties and inability to creation of chelates. Humic acids are strongly connected with mineral portion of soil in this type of humus.

Third type of humus is characteristic for semidesert soils. In this humus predominate fulvic acids fraction, whereas arise of humic acids is limited. Beyond this, humic acids are largely bounded with mineral portion of soil.

HUMUS CONTENT OF SOIL

Humus content in soils fluctuating in broad range.

On humus content have influence the following factors :

- Amount and quality of humus, which get at soil in given bioecological zone
- Tempo of humification process of organic matter
- Tempo of mineralisation of humus, which is contain in soil
- Chemical, physico-chemical and physical soil properties
- Amount and quality of mineral compounds contained in soil

Table. Humus Content in Accumulation Horizons of The Main Soil Units in Poland

Division and order	Type, genera and kind	Humus content per cent
Calcisols :	Calcarious Rendzinas	3.4 2.1–6.3
	Jurasic Rendzinas	4.4 1.5–7.0
Phaeozems	Haplic Phaeozems	2.8 1.8–4.0
Cambisols :	Brown soils formed from sands	1.5 0.9–2.2
	Brown soils formed from light and medium loams	1.8 1.1–3.0
	Brown soils formed from heavy loam	2.5 1.6–3.7
	Brown soils formed from silt formations	1.7 1.3–1.9
	Brown soils formed from loess and loess-like materials	1.9 1.4–2.6
Luvisols :	Grey brown soils formed from silt formations	1.9 1.4–2.4
	Grey brown soils formed from loess and loess-like materials	1.8 1.0–2.5
	Grey brown soils formed from light loam	1.6 1.0–2.6
Podzols	Podzolic soil formed from sands	1.5 1.1–2.0
Gleysols :	Boggy soils formed from silts	1.6 1.2–2.1
Gleysols :	Black earth formed from sands	2.8 1.2–4.1
	Black earth formed from light and medium loams	2.6 1.2–5.7
	Black earth formed from heavy loams and clays	4.9 2.5–5.6
Fluvisols :	Alluvial soils formed from sands	2.9 1.5–5.2
	Alluvial soils formed from silts	3.5 1.7–5.8
	Alluvial soils formed from clays	4.2 2.4–6.8

Chapter 24

Indigenous Soil and Water Conservation

INTRODUCTION

From 1997 to 2001, the second phase of the Indigenous Soil and Water Conservation (ISWC 2) programme had been operating in seven countries in both Anglophone (Ethiopia, Tanzania and Zimbabwe) and Francophone Africa (Tunisia, Burkina Faso and Cameroon). The programmes had taken an approach to agricultural research and development through building on local innovation and encouraging farmer experimentation and Participatory Technology Development. This paper describes the main components of this farmer innovation approach.

Major Components of the Farmer Modernism Approach

This presentation of ten components of a farmer innovation approach to agricultural research and development does not mean that these were or should be neatly implemented one after the other. In practice, several components run parallel to each other or are repeated several times within the approach. Nor does it mean that each country involved in ISWC 2 and PFI has given the same attention to each component. The vastly differing socio-economic and agro-ecological conditions in each of the countries involved have led to different accents being placed. The three country programmes under PFI followed the steps fairly closely, whereas each of the seven countries under ISWC 2 designed its programme in the light of its own perceptions, experience and history, as well as the capacity and orientation of the partner organisations.

Members of the ISWC 2 European Consortium gave some advisory support or 'backstopping', but it was up to the National Co-ordinator and Steering Committee in each country to determine how they would combine the components of the farmer innovation approach to suit their own circumstances. During the annual review meeting and via the joint ISWC/PFI newsletter on farmer innovation in land husbandry, the partners in the country programmes could compare and analyse their different experiences and learn from each other. Also the backstopping missions following Consortium meetings and the sharing of reports between country programmes contributed to this exchange.

Within each country, the ISWC 2 and PFI programmes have concentrated their initial work in selected regions, for example, the Southern Highlands in Tanzania, parts of the Central Plateau of Burkina Faso, the Tigray Region in northern Ethiopia, the Kabale District in south-west Uganda. An exception has been the Zimbabwe programme, which already had considerable experience in participatory research and extension in Masvingo Province before the programme began and used the ISWC 2 phase for scaling up. Most of the components described below refer only to activities in the action areas, but the lobbying for policy change and institutionalising the farmer innovation approach (Components 9 and 10) have reached out beyond the action areas, with a view to further expansion of the activities.

Training in PRA and PTD

Although there are outstanding exceptions, many scientists and extension agents have difficulties to communicate eye-to-eye with farmers. Their conventional training and vision of development through transfer of modern 'improved' technologies have imbued them with a sense of superiority. They tend to regard themselves as the teachers of farmers, as the agents of change to draw farmers out of the stagnancy of tradition. Most of them cannot imagine that they can also learn from (illiterate) farmers.

With a view to changing these attitudes, the ISWC 2 programme organised training-of-trainer workshops in PRA and PTD. Two courses were held on a regional level for the anglophone and francophone countries in Zimbabwe and Burkina Faso, respectively. The trainees from each country programme came from research, extension and development organisations, both governmental and non-governmental. They then collaborated in planning and facilitating PRA/PTD training and awareness-raising workshops at national and/or provincial (regional) level. Some countries have invested more time in PRA/PTD training than others. For example, ISWC-Ethiopia built up a local training capacity through a series of workshops and by including new people as co-facilitators of training workshops. It was thus able to train numerous government extension agents (called Development Agents or DAs) in all four regions of Tigray. The philosophy behind this was to create a broad basis for identifying local innovations and for supporting farmer-led experimentation and farmer-to-farmer exchange, rather than starting a few isolated PTD experiments involving only a small number of farmers and DAs and then trying to scale up to include a larger number.

Likewise, in Tanzania, a major output of the training activities has been the firm establishment of local PRA/PTD training capacities : already ten professionals from research, extension and training organisations are well practised in facilitating participatory training events. ISWC-Tanzania developed a sequence of training events focused mainly on scientists, extension staff and farmers directly involved in the programme. An introductory PRA/PTD training was followed by a refresher training to review experiences and increase insights. This, in turn, was followed by training in farmer experimentation and by review and planning workshops based on the experiences in farmer experimentation. Thus, training and learning have been closely integrated with exchange, joint review and planning, with the latter gradually assuming more importance than training *per se*.

Because one external training course is not enough to impart all the necessary skills, most of the ISWC National Co-ordinators asked the external backstoppers to support them in implementing follow-up sessions at national level. These usually focused on specific aspects of PTD, such as how to design experiments together with farmers. In Cameroon, the backstopper was directly involved in facilitating a workshop to raise awareness about PTD at national level. However, in most countries, the PRA/PTD training events at national level were carried out by the participants in the initial training-of-trainer workshops and other local facilitators, with the backstoppers giving advisory support often only from a distance through electronic means of communication.

In the case of ISWC-Tunisia, the scientists had little experience with participatory research and were hesitant to facilitate the training on their own. For the first national PRA/PTD training workshop held in southern Tunisia in October 1997, trainers were flown in from Senegal and Burkina Faso – an innovation in itself, in the sense that experts from South of the Sahara were training scientists in North Africa. The workshop was an eye-opener for most of the participants, who were astounded at the level of local knowledge they discovered during the fieldwork. After this workshop, however, only a small number of participants continued to use PRA techniques in their regular work. The reasons have not yet been analysed, but it is likely that isolated trainees could not easily apply their new knowledge and skills in an unchanged institutional setting. The few trainees who did continue to use their new skills managed to identify interesting innovations and innovators, and this had a decisive positive impact on the future of the ISWC programme in Tunisia.

Not only in Tunisia but across all seven countries in the ISWC programme, the success of the training events in changing the participants' attitudes depended on the personalities of the individuals and the extent to which the institutional contexts in which they operate gave them space to use participatory approaches. The National Co-ordinators and external backstoppers observed in many cases that, when scientists and extension agents were interacting directly with farmers in the field in post-workshop activities, they treated the farmers with greater respect than previously. In some cases, they gave the innovators nicknames, such as 'professor', to express admiration for the farmers' wealth of knowledge, experience and creativity, as well as for their capacity to express what they know and do. Indeed, it became evident during field days and travelling seminars that some farmer innovators have oratory and presentational skills far superior to those of many scientists, college lecturers and professors.

Identifying and Verifying Farmer Innovators

The process of identifying farmer innovators is not an easy and straightforward one, because farmers are not necessarily aware that they are innovating. In each country, all partners first had to agree on the concept of innovation and innovators. For example, as a working definition for an innovator, ISWC-Ethiopia used 'someone who develops or tries out new ideas without the support of formal extension services.' 'New' was defined as 'something that has been started within the lifetime of the farmers - not something that s/he has inherited from parents or grandparents'. This definition helped to focus the search on the dynamics of indigenous knowledge, not just looking at what is traditional but seeking endogenous change. In contrast, ISWC-Tunisia decided to include also technologies inherited from parents in the inventory of local innovations in land husbandry.

Many types of actors were involved in the identification of farmer innovators : they included extension agents, NGO field staff, students doing fieldwork, teaching staff and scientists. In the search, the detailed local knowledge of field agents in NGOs and government services was an important asset. They could sometimes name innovators whom they already knew from their day-to-day work. However, this was not always the case. Some field agents initially had eyes only for the technologies that they themselves were promoting and for the 'progressive' farmers who adopted these technologies. For example, NGO staff members in Cameroon were surprised to hear about the existence of farmer innovators; one of them said : 'We are always in the field and we have never seen them.' However, once they became familiar with the concept of farmer innovation and had actually met some local innovators and seen their accomplishments, the same staff members began to recognise innovators (both men and women) regularly. Some of them even commented that they were now noticing small differences in farming practice almost every time they went to the field.

Not only a wide range of actors was involved in the identification process; also a wide range of means was used : direct observation, interviews with farmer groups and with key informants, contests, peer identification, invitations for self-identification through rural and regional radio programmes. In Tunisia, for example, men and women farmers responded to the weekly radio programme on Agriculture and Innovation by writing to the broadcasting station that they, too, had innovated and wanted to tell others about this.

Some methods worked well in some countries but not in others, for readily understandable reasons. For example, farmers in Ethiopia could readily identify innovators in their midst but when innovative farmers in Burkina Faso were asked to identify other innovators, they tended to name relatives. This was a logical thing to do, because ISWC-Burkina provided credit to innovators. In Ethiopia, an annual contest was set up : village elders, local leaders and DAs identified top innovators in each district. The winners were distinguished in public ceremonies, as an expression of recognition by their fellows and to encourage them to continue innovating and interacting with their communities. ISWC-Ethiopia deliberately avoided offering credit to individuals, in view of the DAs' negative experiences in having to convince farmers under the Sasakawa Global 2000 scheme to take up credit for purchasing external inputs. This illustrates how each country designed its programme in the light of its own experience.

One issue that was much discussed by the partners in the ISWC 2 programme was whether the focus should be on individuals or on groups of farmers in identifying and giving public recognition to innovators. Most countries initially focused on identifying outstanding individuals. There was sometimes a tendency for scientists and extension agents to select those farmers with whom they were already in contact. Other countries tried to spread the search as widely as possible, relying mainly on extension agents and village leaders but rarely involving 'ordinary' farmers in the identification process. In its work in Tigray, the ISWC-Ethiopia team encouraged partners to identify innovations by both individuals and groups of farmers (*e.g.*, in modifying traditional irrigation systems), as well as innovations by husband-wife teams. However, in most countries, it was often only later, after starting to work with male innovator in participatory research, that the important role of wives in local innovation and experimentation was realised.

In both ISWC 2 and PFI, an innovation was generally defined as something new to the particular locality, but not necessarily new to the world. For example, an Ethiopian farmer who saw a *shadouf* (an ancient Egyptian device for lifting water) while he was a refugee in the Sudan and who developed a *shadouf*-based irrigation system when he returned to Tigray, was regarded as an innovator in his home area. In the PFI programme, farmers who simply copied what another farmer in the same village had developed or introduced were not regarded as 'first generation' innovators, even though what they did was new for their particular farms. A first round of identifying farmer innovators by the PFI programme therefore had to be followed by a process of verifying that they were indeed 'first generation' innovators in the eyes of their fellow villagers.

In most countries in the ISWC 2 programme, the distinction between 'first-generation' and 'second-generation' innovators (adopters, adapters) was not made. The focus was on identifying new ideas that had been developed or tried by a few farmers but could be of interest to many : in other words, the programmes were more concerned with verifying the usefulness of the innovation rather than whether a particular individual was a genuine innovator. In order to assess whether an innovation was likely to be more widely accepted, some countries, such as Ethiopia, Cameroon and Tanzania, studied the social setting and resource endowments of the innovators.

A total of over 800 farmer innovators have been identified in the seven countries involved in ISWC 2 and the innovations of most of them have been documented in reports, database entries and photographs. In some cases, documentation had taken the form of video films. Not only scientists but also field agents have been preparing this documentation. These few hundred innovators represent only 'the tip of the iceberg', because the programme covers relatively small parts of each country. Most innovators and their innovations continue to be hidden, overlooked or ignored. The main aim of the programme is not, however, to identify only outstanding innovators. It is to stimulate the innovative capacities of *all* farmers, by encouraging as many scientists and extension agents as possible to recognise local innovation and to work with farmers in ways that promote innovation, as outlined in the following components. In this context, it is important that farmer innovations are 'verified' by other farmers as part of a process of stimulating more widespread innovation in a village.

Analysis of Innovators and Innovations

Because relatively little is known or documented about farmer innovation in Africa, the ISWC 2 and PFI programmes tried to gain a better picture of the characteristics of outstanding innovators and their motivations for innovating. As pointed out by Yohannes, every farmer has to be an innovator to some degree, but some farmers are well known in the community as being particularly innovative. What triggers these farmers to develop new ideas or try out something new, while others operating under similar conditions do not? Is it because they are particularly rich or particularly poor in productive resources? Are most of the outstanding innovators old or young, educated or illiterate, men or women? Do innovations by women differ from those by men? Do the more innovative farmers have certain characteristics that can be cultivated, *e.g.*, through education and training? In which domains do farmers innovate? What is the potential of local innovations to be adopted or adapted by other farmers?

The viewpoint of the ISWC 2 and PFI programmes from the viewpoint of doctoral research on the subject. Generally, the men and women who were found to be outstanding innovators are people with strong personalities, who have sometimes had to withstand initial resistance by other villagers. This was particularly the case among the women innovators in Tigray who dared to challenge cultural taboos by taking on the male task of ploughing or using donkeys for traction. Some also had to deal with the resistance of family members. For example, a man in Tigray who spent several years developing an intricate system of lifting, storing and distributing water to irrigated plots was threatened with divorce by his wife, because her family thought he was wasting his time on the farm instead of going to town to earn money in construction work. Only after he started to sell fruit, vegetables and fodder from the irrigated land did his wife return to live with him. She now collaborates closely with him in maintaining the irrigated plots and selling the produce.

In Tanzania, a systematic study was made into the socio-economic background of the farmer innovators identified through the programme activities. Students from Wageningen Agricultural University found that the innovators included both rich and poor farmers. The former appeared to be innovating because they are curious and enterprising and had the resources to experiment; the latter appeared to be innovating because they were highly motivated to improve their situation and, having no resources to apply solutions offered from 'outside' (the extension service), used their creativity to find their own solutions.

Discussions among partner organisations in the two programmes often revolved around the question whether there was a tendency to work primarily with richer farmers who had developed more spectacular innovations, or whether outstanding innovators were richer because they were so innovative. There was a concern that the smaller improvements in their farming systems made by farmers with fewer resources were being overlooked. When country programmes such as Tunisia and Ethiopia began to seek out especially women's innovations, they found that these tended to be low-cost, making use primarily of local resources. Such innovations are particularly promising for resource-poor farmers, not only women, and have a good potential for spreading, but these innovators usually do not have the same confidence and means to make their ideas more widely known as do richer, male farmers. Extension agents can play a particularly important role in encouraging farmer-to-farmer communication about the low-external-inputs options developed by the less wealthy local innovators.

Setting up Monitoring and Evaluation Systems

When all is said and done, it must be possible to show all stakeholders in agricultural development whether or not the programme has enhanced the capacity of farmers to experiment and to innovate, and whether this increased capacity and the emerging innovations indeed contribute to improving natural resource management and the well-being of the farm families. It is important to produce numbers, especially for donor agencies and political decision-makers, but it is just as important to analyse the processes, as a source of learning and guidance for future work. It is therefore necessary to document *who* has done *what, where, how* and *why*. If monitoring and evaluation (M&E) of the programme activities is to be done well and be useful, then all partners must be involved. They must feel both capable and willing to do the work. Therefore, the observations made and recorded have to be limited to the smallest possible number of key indicators of interest to those involved, and the questions need to be readily understandable. Only in this way can the learning aims be achieved.

Members of the ISWC 2 Consortium proposed questions and forms to facilitate monitoring of the process and outcome of the various programme activities, as well as to assess their impact. The proposals included more conventional aspects of M&E – collecting information defined by outsiders for analysis by outsiders – and some participatory aspects of M&E, so that the farmers and national partners could record and assess according to their own criteria and using methods they chose or developed themselves. It was up to each country programme to adapt the questions and forms to the local conditions as well as to the local capacities to handle and analyse the information generated.

An initial assumption of the ISWC 2 Consortium was that farmers would perceive the recording of data as a burden. However, the experience in, for instance, Burkina Faso, Cameroon and Uganda has been that, when the farmers were involved in determining what data to collect, they were keen to keep the records. The farmers regarded this as a new and useful skill that helps them manage their farms better, because recording allows them to keep track of and analyse levels of inputs, outputs and crop performance. All the farmer innovators involved in joint experimentation in the Yatenga region of Burkina Faso insisted on keeping the records themselves. After one season, one innovator remarked : '*On ne mange même plus au hasard*' (even our meals are no longer unplanned). Some farmers in Uganda seemed to be more interested in improving their capacities to keep their own records than in improving their yields through the particular technologies they were testing. Many farmer innovators also keep visitors books, so that they can record who visited them over the years.

Exchange Visits and Study Tours for Farmer Innovators

Both the ISWC 2 and the PFI programmes gave innovators an opportunity to visit each other's farms and to exchange information and experience about their innovations. After their return, many innovators started to try out for themselves what they had observed.

Farmer exchange and study visits have been highly effective in disseminating local innovations or the principle ideas behind them. The major question is how to widen the impact of this farmer-to-farmer communication. Three phases can be distinguished in the process : preparation, implementation and follow-up. In exchange visits in general, and this is also true for some of the exchange visits under ISWC 2, most attention is given to implementation and little to adequate preparation of both the visitors and the hosts. What do both sides hope and expect to gain from these visits? Also an adequate follow-up is important. What did the visitors and hosts find useful about the exchange? How and to whom did they report their impressions? Did the visitors adapt the techniques to their specific conditions back home? Some tentative observations in Zimbabwe are that women seem to be more likely than men to share their impressions with others, although not necessarily in a formal way, and that women tend to share with women while men tend to share with men.

It is also important to know whether some additional technical training or other forms of support may be required as a follow-up to a visit or study tour. In Tunisia, some farmer innovators have trained other farmers, *e.g.*, in grafting fruit trees, immediately during the visits, but often such training will be possible only after the actual farmer-to-farmer visit.

In the PFI programme, innovators are encouraged to organise themselves into clusters of usually eight farmers, and they exchange experience within and between these clusters. One advantage of such small groups in one locality is that the members can meet easily and at little expense, at least in the case of groups in more densely populated areas. The ISWC programme in Uganda follows the PFI approach and includes 32 farmer innovators in four networks of eight innovators each.

In the other ISWC 2 countries, it was left up to the innovators to decide if and how they wanted to organise themselves. In Ethiopia, participants in the first travelling seminar decided to form an association and to meet annually to discuss progress in their innovation activities. In contrast, the farmer innovators involved in joint experimentation with scientists in Cameroon were selected by their peers within an existing farmers union and, in Zimbabwe, they were selected by self-selected groups of farmers with common interests. In Tanzania, three cases were recorded in which farmer innovators, after returning home from exchange visits, started up a local group of one or more innovators and a large number of neighbouring farmers. They not only wanted to share what they had learn about techniques; they also found that participation in the visits had encouraged them to undertake joint development activities.

Farmers' Evaluation of Local Innovations

This component can be a preparatory step to the selection of promising ideas for farmer-led experimentation as well as to farmer-to-farmer exchange on innovations. In most cases, innovations were initially assessed by outsiders. Then the farmers themselves - both the outstanding innovators and their neighbours - were given an opportunity to assess the innovations jointly. This was not limited to comparing yields and other technical parameters. For example, Yohannes found the values and experience of the communities strongly influenced their acceptance of local innovations.

Village workshops have proved to be a useful tool for farmers' assessment of innovations. In Tigray, for example, the Bureau of Agriculture and Natural Resources (BoANR) organises Farmers' Fora during which farmer innovators and their neighbours come together to examine both local and external innovations and to discuss what is useful for which categories of farmers in the community. Such workshops for farmers at village or district level can also help to counteract the isolation from their fellow villagers that many outstanding innovators feel because they are doing something different from the norm.

In Zimbabwe, for example, local innovators were often regarded - at least by government officials - as 'off-types', because the policy environment discouraged the application of indigenous knowledge. The emphasis of ISWC-Zimbabwe on group formation and joint experimentation served to enhance the social standing and confidence of innovators and strengthen their position *vis-à-vis* the government services. A group approach also reduces the risk that farmer innovation programmes focus on only a few privileged farmers.

Innovation assessment workshops at village level can be starting points for designing experiments implemented by farmers who have been selected by the community. This embeds the experimentation by individuals within a wider social process. In Cameroon, for instance, the priorities for experimentation were set during a workshop involving scientists, extension agents, farmer innovators, other male and female villagers, representatives of the traditional authorities and, for some time, the *Fon* (the king) himself. This was deliberately done to avoid isolation of the innovators and their experiments from the rest of the community.

Stimulating and Supporting Processes of Joint Experimentation

The original assumption in the ISWC 2 programme document was that joint experimentation by scientists, extension agents and farmer innovators. This was far too optimistic. The entire process of gaining a common understanding of concepts, training in PRA/PTD, identifying and understanding farmers' innovations and informal experiments, organising exchange visits and building up partnerships between farmers, extension agents and scientists requires much more time.

If the joint experimentation had started earlier, there would have been a greater risk that the scientists defined the experiments and imposed them on the farmers. Ideally, the innovators and other interested farmers should set the research agenda and it should be based on their priorities.

Scientists have a role in :

- Proposing elements for testing, which the farmers might like to include;
- Advising farmers how to design simple experiments, so that both farmers and scientists have a firmer basis for evaluating the results; and
- Explaining reasons for the farmers' findings and thus helping farmers better understand some of the principles and less visible factors influencing the outcome of their experiments.

In addition, the scientists can assist in generating 'hard' data to validate the findings in conventional scientific terms, in order to convince other scientists, policy-makers and donor agencies. A further role of scientists, especially in the initial stages of collaboration with farmers and extension workers, is to observe and analyse the process in order to develop the farmer innovation methodology and to describe it clearly to others wanting to engage in it.

The actual process of setting the joint research agenda differed from country to country. In Cameroon, for instance, village workshops were held to elicit the priorities of farmer innovators. One innovator, Samuel Toh, clearly formulated their priorities and told the scientists : 'We are aware that the researchers have their priorities ... Let's first address our priorities and then we will also work with the researchers on what they want to study.' The scientists agreed and a process of PTD was commenced. The situation was quite different in Tunisia. Here, the National Co-ordinator is a scientist who has developed both simple and complex, but effective, technologies that reduce maintenance requirements of a traditional water-harvesting system and allow more efficient use of the limited water resources available in the region. He strongly influenced the research agenda, but this was based on farmers' practice. One farmer innovator in Tunisia observed the scientist's experiments and started his own, in a kind of competition, claiming : 'What you can do, I can do better and cheaper'. It was not initially a strategy of ISWC-Tunisia to set up competition between formal and informal research, but this served to spur on both the scientists and the farmer innovators.

In Tanzania, a series of workshops was organised for the 'experts' (scientists and extension agents) to bring them to acknowledge and appreciate the potentials and important roles of farmers in agricultural research and development. At the same time, workshops and exchange visits were organised for farmers to build up their confidence and assertiveness, so that they would be better able to articulate their position and argue for their interests. Once these two parallel processes had matured, the experts and farmers were brought together in a workshop to set priorities for joint experimentation.

At the start of the ISWC 2 programme, the country with the most experience in stimulating and supporting processes of joint experimentation was Zimbabwe. Here, the PTD activities were inextricably linked with the development of local institutions and were carried out with groups of farmers defined by the communities. PTD thus became a tool not only for developing techniques but also for developing communities.

In some cases, the joint experimentation focused primarily on scientific validation on the local innovation. Here, the procedures were more conventional and it was important to collect data that could be subjected to statistical analysis. In other cases, the process of experimentation was designed to allow primarily farmers to assess the local innovations and/or to investigate farmers' and scientists' ideas to improve them still further. This had the additional purpose of strengthening local capacities to test and assess new ideas. Here, it was more important that the procedures and measurements could be managed completely by farmers, also eventually on their own, and it also meant that farmers were given the chance to learn about experimentation by analysing their own experiences and mistakes. The workshops for farmer experimenters organised by ISWC-Tanzania reflect this latter approach : with the support of scientists and extension agents, the farmers examined each other's experiments and themselves recognised where improvements should be made. In Tanzania, after a small number of joint experiments by farmers and technical specialists in 1997/98, more than 30 joint experiments on a wide variety of topics were carried out in the following two growing seasons. In addition, some individual farmers did their own experiments on the same topics, without direct support from the specialists.

The studies on the maise-pit system are showing astounding results, which are now being acknowledged by the scientific community. Results achieved in farmers' trials with irrigation to induce flowing in coffee are also very promising, and have led to new research questions needing attention, particularly with respect to soil fertility. At the same time, partners in the ISWC-Tanzania programme see a need to stimulate continued experimentation and innovation by farmers, in addition to this more systematic joint experimentation. For example, while the joint experiments with the maise pit system was confined to an agreed set of dimensions for the pits, several farmers have independently tried out other dimensions and ways of making the pits. Discussion between farmers and scientists about the reasons for these changes and the results obtained could enrich the learning process.

Farmer-to-Farmer Dissemination of Innovations

A key aim of both ISWC 2 and PFI is to disseminate useful innovations. Various forms of farmer-to-farmer communication are facilitated, primarily through visits and workshops. The PFI programme organises three types of exchange visits: farmer innovators visit other innovators to learn from each other (FIs to FIs), 'ordinary' farmers visit farmer innovators to learn from their experience (Fs to FIs), and farmer innovators visit other farmers to inform them and to train them in the use of improved technologies (FIs to Fs). The first type, involving only farmer innovators, comes under Component 5, while the second and third types are examples of wider farmer-to-farmer extension.

In the ISWC 2 programme, each country has developed its own variations. In Ethiopia, Components 5 and 8 were combined in the travelling seminars, during which innovators visited other innovators and, at each site of innovation, met with a wider group of people in that community. In Zimbabwe, mixed groups of farmer innovators and other villagers visited other sites of innovation, and the villagers rather than the programme chose the farmers to be involved. In Uganda, study tours were also organised to take farmers to other parts of the country : exchange visits were made between the farmer innovators working with the PFI programme in the north and those working with the ISWC programme in the south. A large number of farmers have been involved : in Kenya alone, over 2500 farmers (1600 men and 900 women) have visited farmer innovators in the framework of PFI.

The original idea was to disseminate farmers' innovations that had been scientifically validated, but it soon became obvious that farmers do not wait until a technology has been tested and given a stamp of approval by scientists. A farmer who thinks that a technology is interesting and possible to apply will try it out, and may adapt it, or may choose some elements to adopt. In Tanzania, seeds and planting materials were exchanged informally during farmer-to-farmer visits and subsequently planted. Some of the local innovations viewed by other farmers have already been widely adopted. Follow-up enquiries after an exchange visit revealed that 79 farmers in Njombe District alone had started trying out the sowing of many maise plants in a pit, an innovation developed by a farmer in Njombe, and also farmers in other districts and regions had started using this technique, although scientists had not yet validated it.

In Tanzania, this component is closely linked with Component 6. The village workshops for evaluating local innovations automatically led to sharing of ideas between farmers. However, most of this sharing probably took place outside of the activities supported directly by the programme. In additional to informal communication between farmers, this included the increased opportunities for farmer-to-farmer exchange within the regular research and extension activities of the partner organisations. For instance, during the regular farmer field days, farmer innovators are asked to explain what they are doing. Information about local innovations and innovators has also been spread via the existing journal of the farmer network Mviwata.

Reactions of different categories of farmers to local innovations still need more investigation. The impact study in Uganda found that women were more responsive than men to adopting new ideas and those who tried out local innovations tended to be poorer in resources than the original innovators. The reasons behind this are not yet clear. It may be that the wealthier farmers are in a better position to take the risks involved in innovating, while the poorer ones are prepared to adjust their practices when they see good alternatives that suit their needs. However, this will apply only where the richer farmers' innovations can also be realised with the limited resources available to the poorer ones. For example, it is unlikely that complex innovations that require high investments, such as those developed by some male farmers in Tunisia, will spread easily through farmer-to-farmer extension without external financial support. Men in Tunisia can take advantage of government subsidies and credit schemes that are not available to most farmers, male or female, in the other countries in the ISWC 2 and PFI programmes.

It is quite a different case in West Cameroon, where – after prices for the men's customary cash crops (coffee and cocoa) fell – some men started innovating with cost-effective ways of growing women's crops. Barthélemy Djambou is one innovator who is exploring better ways to grow vegetables as market crops, and his innovations are of interest mainly to resource-poor women, who customarily grow food crops and vegetables. Here, too, it has yet to be investigated whether women were also innovating with these garden crops and, if not, what was constraining them from doing so.

A fundamental issue that will determine the success of farmer-to-farmer dissemination of local innovations is whether innovators are willing to share their knowledge and experience with others. A woman innovator in Tunisia had not been willing to tell other farmers about how she uses decomposing manure to hatch eggs, until she was offered an opportunity to present her innovation on the radio. Some individuals are not at all willing to inform others about their innovations, and some will do so only for payment (farmers can hire their services as consultants). Nielsen found that the most innovative farmers were the least keen to share their knowledge.

In most cases, however, the ISWC 2 and PFI programmes found that farmer innovators were eager to promote their own innovations. In some cases, they did so because this gave them public recognition and social esteem; in others, because they felt it was their duty to their community. Some farmers deliberately share their knowledge with others in their village in order to avoid jealousy and possible ensuing difficulties (witchcraft).

Women whose innovations go against social norms are initially reluctant to share their ideas openly. The poorer farmers, especially the female household heads, often find it difficult to spare the time to teach other farmers. In one case in Ethiopia, the community decided to send someone to work on a woman innovator's farm when she was asked to teach other farmers, *e.g.*, to train other women how to plough in their own fields. In areas like Tunisia, where the communication radius of women is confined, radio may be the best means for them to share beyond their immediate family and female neighbours.

Farmers share their knowledge informally in various ways, but the most common forms appear to be :

- Farmers discuss new ideas during market visits or at small social meetings, *e.g.*, Ethiopian women who meet for the traditional coffee ceremony;
- Farmers visit other farmers who are doing something that appears to be interesting; and
- Farmers speak to a larger group about what they have seen or are doing, *e.g.*, during a village meeting, possibly held for another main purpose.

In Burkina Faso, outstanding personalities have innovated in developing extension approaches to share experiences with improved traditional planting pits (*zaï*). One innovator initiated an annual *zaï* market day, which is attended by all farmers interested in exchanging information about this technique and its numerous variations. Another has created a kind of 'field school' : farmers jointly treat a piece of degraded land with *zaï* and learn by working together. A third innovator coaches a number of other farmers, one of whom has started to experiment with the layout and configurations of *zaï*. These models retain many of the characteristics of informal learning between farmers : social gathering, learning by seeing and doing, and personal mentoring. The value of dialogue between recognised equals and personally experiencing new ways of working, rather than being told how something should be done, is expressed in the farmer's statement : 'Learning from exchange visits is better than being visited by a village extension worker'. While both ISWC 2 and PFI have placed great emphasis on facilitating farmer-to-farmer exchange, designed to reinforce the above-mentioned informal processes, PFI has taken a more structured approach to this.

Raising Awareness and Lobbying for Policy Change

The eight countries participating in the two programmes have pursued various paths to raising awareness about farmer innovation and trying to influence policy in its favour. Much attention has been given to documentation and publication in the form of working papers, project reports, workshop proceedings, papers for conferences and articles in newsletters and journals. The programmes in Cameroon and Ethiopia even started up their own local newsletters. All the ISWC 2 teams have gained access to radio, television and the press. As soon as ISWC-Ethiopia had identified some outstanding farmer innovators, it arranged radio interviews that were broadcast in the local language (Tigrigna). ISWC-Tunisia launched a two-hour programme on Agriculture and Innovation, which is broadcast by a regional radio station to most of central and southern Tunisia. The Tunisia experience stimulated ISWC-Cameroon to appoint a media advisor to maximise access to mass media and stimulated ISWC-Burkina to start, in late 2000, a radio programme in the Yatenga region using the station *La Voix du Paysan* (Voice of the Farmer). Such radio programmes are likely to be noticed also by decision-makers at regional and national level; however, the impact in this respect has yet to be investigated.

Several ISWC 2 teams have produced one or more video films about farmer innovation in their country and, in some cases, at least excerpts have been shown on national television. PFI has produced a broadcast-quality video on farmer innovation in East Africa, which has been aired on television in all three countries involved in that programme. It has also published a book outlining the theory and practice of its approach and more than half of the 10,000 copies had been distributed by the end of 2000.

Various other activities also contribute to enhancing policy dialogue. The National Co-ordinators have organised meetings with and field visits by policy-makers. The President and Vice-President of Uganda visited farmer innovators under the PFI programme. In most other countries, meetings have been held with Ministers of Agriculture or with senior staff responsible for national programmes of research and extension. Upon the invitation of the country programmes, these policy-makers opened national or international workshops on farmer innovation.

The Head of the BoANR in Tigray plays a very active role in the Steering Committee of ISWC-Ethiopia. The national ministers of both Agriculture and Education have visited farmer innovators in the field. The Integrated Food Security Desk (IFSD) in Tigray recently decided to introduce the approach of promoting farmer innovation and farmer-to-farmer extension in the 16 most drought-prone districts. According to Mengesha Haile, the Head of the IFSD, the performance of the DAs will no longer be evaluated according to how many farmers they draw into the Global 2000 scheme, which may not be appropriate for drought-prone areas.

In trying to initiate an approach to PTD through identifying farmer innovation, ISWC 2 deliberately commenced work below the national level (*i.e.*, in only certain provinces or regions of a country) in order to develop successful cases of proven local innovation capacity, as this is essential for effective lobbying work. Within three years, the country programmes collected substantial evidence that the farmer innovation approach produces results of interest to small-holders and supporting organisations, and that it can be an alternative to the conventional ToT model of agricultural research and extension. They are providing the accumulated evidence in various forms to decision-makers in their countries, and are exposing them to examples of farmer-led experimentation and innovation.

Important events for raising awareness about the farmer innovation approach on national and regional levels were the two conferences on farmer innovation held in November 1999 in Cameroon for francophone (primarily West) Africa and in February 2000 in Ethiopia for anglophone (primarily Eastern and Southern) Africa. These meetings gave an opportunity for exchange of experience between the countries in the two programmes, but also broadened the participation to include other projects and, above all, policy-makers who could have a positive influence on institutionalising the farmer innovation approach. Also for this purpose, one day of the annual review meeting in the third and fourth years of the ISWC 2 programme was opened up to policy-makers from the host country.

Institutionalising the Farmer Innovation Approach

Ultimately, the ISWC 2 and PFI programmes are working towards incorporating the farmer innovation approach into the regular activities of agricultural research, extension and education in the countries where they are operating. Only in this way will it be possible to ensure the long-term sustainability of the approach. This requires a shift away from the conventional ToT towards the promotion of farmer-led experimentation based on both indigenous and exogenous ideas. Both programmes seek to strengthen the capacity of scientists and extension agents to recognise the problems being addressed by local innovators in their informal experimentation and to participate in and support farmer-led research, as outlined in Component 7. In this collaboration, one role of extension agents will still be to make information, from whatever sources, more easily available to farmers. However, this will be with the aim of enriching farmers' experimentation, *i.e.*, feeding farmers' hunger for useful new ideas to try out, rather than prescribing introduced 'improvements' which may not prove to be such.

Virtually all persons who have been directly involved in the two programmes have become excited about the approach. Especially the scientists and extension agents involved at field level became enthusiastic when they started to recognise the knowledge and skills of local innovators. Both programmes have contributed to building up national capacities to give training in PRA and PTD. This can form a basis for further expansion of activities within countries currently in these programmes and to new countries in Africa and elsewhere.

However, real progress and substantial scaling up can be achieved only when the methods of promoting farmer innovation are included in the curricula of research, education and training institutions. Mekelle University in Tigray, the lead agency of ISWC-Ethiopia, has started to incorporate PRA/PTD training into its curricula; so has the Cooperative College Moshi, the lead agency in Tanzania. Some students of Mekelle University have already done their practical attachments on indigenous knowledge, farmer innovation and informal experimentation; and scientists have become involved in studies related to farmer innovation. ISWC-Cameroon has recently trained staff of the National Agricultural Extension and Research Programme (PNVRA) funded by the World Bank and IFAD and, in the next phase, the farmer innovation programme will become a semi-autonomous project under the PNVRA. The involvement of Masters students from Wageningen Agricultural University in the Netherlands has led to an increased understanding of and interest in the farmer innovation approach among staff of the Department for Communication and Innovation Studies. These are encouraging steps forward, but there is still a long road to travel to institutionalise the farmer innovation approach in these and other countries in Africa and Europe and to make the approach less dependent on external funding and technical support.

MEANING AND SCOPE OF WATER CONSERVATION

Water conservation is a broad term involving all the measures which aim at arresting maximum amount of run-off water, storing or collecting it in safe, artificially raised structures and recycling the stored water again to the field in times of need. The ultimate aim of water conservation is to produce maximum from the limited availability of water, by using it most efficiently. An efficient system of collecting, storing and reusing run-off water is sometimes also referred to as water harvesting.

From the point of dry land agriculture, water harvesting indicates not only storing rain water in irrigation tanks/ponds but also employing mechanical and cultural manipulations of soil to improve its rainfall absorption and moisture retention capacity. However, the measures adopted are location specific. To decide on the suitable measures to be taken a preliminary inventory of soil water and other resources, a study of the land use pattern, a dialogue with the target group and a study of socio-political, economic and cultural setting are a must. This implies a scientific approach to water conservation. The need for educating, and organising people to make them accept necessary reforms for effective implementation of integrated water conservation programmes need not be over-emphasised.

The chief agent of soil erosion is water. It is estimated that about 5333 million tonnes of soil (amounting to about 16 tonnes/ha) is eroded from Indian territory. Nearly 29 per cent of this is carried into the sea and nearly 10 per cent is deposited in surface reservoirs resulting in loss of storage capacity of reservoirs and occurrence of floods. Conservation of water virtually reduces the soil erosion. Thus the scope of conservation is very wide, encompassing all the measures of soil conservation, moisture retention and nutrient preservation in the soil.

Need of Water Conservation

Water is absolutely essential for utilisation of all other natural resources including land. Agriculture depends on the availability and supply of water either from rainfall or from surface or underground fresh water resources. Most of the water is received through monsoon rains and is obtained from July to September. Only in some rivers of North India, water is received through the melting of snow in summer.

Because of high intensity of monsoon rains coupled with poor infiltration rates of soils, there is a heavy run-off of water. Run-off occurs when amount of precipitation is more than infiltration of water into the soil. Run-off is very heavy particularly in first rains. As a consequence of heavy run-off, the discharge in rivers increases during rainy period. The experience of last 2-3 decades, shows that every passing year the amount of run-off water discharged into the rivers is in creasing. It indicates that the amount of water which was conserved earlier in the site of rainfall (catchment area) has been reduced.

Our country receives more annual rainfall than the global average of annual rainfall. Therefore, water conservation has to be adopted as a national objective, not only for the sake of water availability but also for protecting the soil which is a precious resource.

Water Erosion Problem in India

Out of the total 328 m ha of geographical area of the country, about 143 m ha is cultivated and about 65 m ha is barren uncultivable waste and cultural waste-lands. By and large these lands are more or less devoid of protective vegetation and physical barriers and hence are subjected to very serious water erosion. Nature takes 500 to 1000 years to develop 2.5 cm top soil but gets eroded in a course of a year or two. As a result it becomes the most harmful single factor in the deterioration of productive land. Along with the loss of run-off water a colossal amount of soil is also lost.

Overgrasing, deforestation, faulty cultivation, shifting cultivation and carelessly built roads in catchment areas have led to devastating down streams. These include gullying and floods leading to destruction of farm lands and villages, drop in flow during the dry season and consequent loss of crops, and siltation of reservoirs and canals. The problem has been further aggravated due to high rate of population growth : both human and livestock, resulting in indiscriminate exploitation of natural sources, for meeting the ever-increasing demand for fodder, fuel, fibre and fertilizers. Thus continuous degradation of production base and imbalance in land-water-plant-man-animal system is leading to ecological imbalance and economic insecurity through severe soil erosion both by water and wind and threat to the quality of our life and cultivation.

Even though no systematic survey for accelerating the extent of water and soil loss in different areas of the country, has been carried out, it is estimated that about 5334 mt of soil is detached annually from its original place. The country's rivers carry 2052 mt of this, nearly 1572 mt is carried away by the rivers into the sea every year and 480 mt is being deposited in various reservoirs.

Soil erosion by water is mainly of four types :

1. *Sheet erosion :* About 4-10 tonnes/ha/year in red soil, 17-43 tonne/ha/year in black soil and 4-14 tonne/ha/year in alluvial soil.
2. *Gully erosion :* About 33 tonnes/ha/year in ravine regions
3. *Hill side erosion :* More than 80 tonnes/ha/year in landslide, mine spoil areas etc.
4. *Stream bank erosion :* Along with soil, about 2.5 mt of nitrogen, 1.5 mt of phosphorus, 2.2 mt of potash, large amount of organic matter and micro-nutrients are also lost. The loss of nutrients is valued at about Rs.15000 million per annum. Mohanjodaro and Harappa are the examples of this fact. Glaring evidence of neglect of rainwater can be seen in the denuded hill tops all around and the extensive network of raville in the states of Gujarat, Madhya Pradesh, Uttar Pradesh and Rajasthan. Once a thick forest, Rajasthan is now a desert, Conservation of water is prerequisite to conserve soil which is a natural reservoir of plant nutrients. Taking into consideration the magnitude of problems, there is a great scope to streamline the entire water conservation programme more scientifically

on micro watershed basis in cluster instead of sporadic efforts. Watershed management programme to conserve water and soil is a new approach and it is getting popularity very rapidly. The programme involves an appropriate combination of agronomic, mechanical, agrostological and afforestation measures, fitting to soil-climatic and topographical conditions of area and socio-economic conditions of the target group.

Types of Water Losses

Water losses with special reference to soil erosion aloe de- scribed below :

Raindrop or Splash Loss

Raindrops fall on the ground and form a splash along with suspended soil patl icles. Water and soil move away from their original place.

Sheet Erosion

When the intensity of rainfall is bit higher, a fairly uniform layer of soil is lost from the soil along with the run-off water. This type of soil and water loss is serious on slightly sloppy lands. The existence of this kind of soil and water erosion is seldom known to farmers as it does not leave behind any visible change on soil topography. Though the field appears the same as before but the cream of the soil *i.e.,* highly fertile upper most layer, is skimmed off along with moving water. The estimates reveal that sheet loss of water erode away many tonnes of soil and has damaged millions of hectares of land in our country.

Rill Erosion

The speed of run-off water is sometimes high enough to create rills on the soil surface. Rills are very shallow finger sised channels which can be smoothened out by ploughing. Rill erosion is the intensified form of sheet erosion and more distinctly noticeable. The width of rills keeps on widening with every passing rainy season.

Gully Erosion

Gully erosion of water and soil is the aggravated form of rill loss. When rills are not smoothened out, water keeps on drawing through them and these shallow rills go deeper and deeper with every passing rainy season. The depth and size of rills go so high that they cannot be smoothened out with ploughing, they are called gullies. The size of gullies may be many metres wide and deep. They provide a big route to run-off water and a huge amount of water is lost through them.

Factors Affecting Water Losses

Water in motion removes the soil and causes erosion of soil. Water erosion is due to dispersive action and transporting power of water. So it descends in the form of rain and leaves the land in the form of run-off. The intensity and speed of run-off are directly correlated to the amount of soil' and water loss. Some of the factors discussed below greatly influence the extent of soil and water loss.

Climate

The major climatic factors affecting run-off and erosion a rainfall, temperature and wind. Of these, rainfall is the most important and the knowledge of rainfall characteristics is important in erosion studies. Raindrop impact on a soil surface has a pronounced effect on erosion. Raindrops cause soil splash, detach soil particles and make them available for transport. The amount of soil detached by the rain depends on the intensity of rain, its duration and the protective value of any cover present.

Erosive capacity of rain drops results from three factors, namely :

1. Amount and intensity of rainfall,
2. The diameter of raindrops, and
3. The velocity of drops as they strike the soil.

The amount of soil transported by raindrops is governed by the transporting capacity of the raindrop energy; the case with which detached particles are called; and the resistance afforded by the vegetation and other obstructions.

The amount, intensity, frequency and duration of rainfall have very definite effects on the amount and rate of resultant run-off. As large total. Rainfall may not cause excessive erosion if the intensity is low. Likewise, an intensive rain of short duration may not produce sufficient run-off to cause erosion. When both amount and intensity factors are high in a given storm, run-off erosion will be serious.

Topography

Sloppy soils frequently permit run-off, resulting in the loss of water and soil, while on flat lands erosion is comparatively negligible. The degree of slope and length of slope are the main features of topography affecting erosion. Velocity of run-off water is influenced mainly by the degree of slope. If the land slope is increased four times, the velocity of water flowing over it is approximately doubled. When the velocity is doubled the erosion capacity as represented by the kinetic energy of flowing water is increased about four times, the amount of material of a given size that can be carried is increased by about 32 times, and the size of particles that can be transported by pushing or rolling is increased by about 64 times. There is a definite limit to the amount of suspended material that can be carried by given stream of water. When this limit is reached, more material cannot be picked up without increasing either the velocity or the depth. Any reduction in velocity on the other hand, will result in deposition of the suspended soil.

Vegetation

Vegetation is one of the most efficient means to conserve water. A thick growth of grass or a dense forest may negate the effect of all those factors which accelerate water erosion.

Vegetation reduces water erosion through the ways mentioned below :

- *Interception of rainfall :* The foliage of vegetation intercepts tile raindrops, retains a portion of water and evaporates it back into the atmosphere. However, the plant absorbs some of this water through foliage and utilises it for metabolic purposes. The amount directly evaporated does not contribute to the run-off water.
- *Decreasing run-off velocity :* Vegetative cover serves as a hindrance to run-off water. A weIl distributed, close-growing vegetation not only slows down the rate at which water travels down the slope, but also tends to prevent " the concentration of water in a particular area. These two effects greatly reduce the erosive capacity of the run-off water.
- *Root effect :* Roots of grasses and other vegetation make the soil porous and permeable to water. The roots die and decay in the soil leaving behind numerous small cavities which increase the pore space and consequently the absorption of water.
- *Biological effect :* Remnants of vegetation serve as the food material for numerous organisms inhabiting the soil. These organisms range from large sised earthworms to microscopic bacteria, fungi, algae and actionmycetes. Macro-organisms work the soil and make it porous and permeable. Micro-organisms decompose organic matter of the soil, and humus is formed as a decomposition product. Humus is very efficient in retaining water. It can retain water as much as 100 times of its own weight.

Soil

The amount of water left on the soil surface for run-off very much depends on soil properties. If the infiltration capacity of the soil is high, a large amount of water is soaked in by the soil and there is little run-off.

In case soil is very compact or hardpan lies in the sub-soil, a large amount of water is lost by run-off. The infiltration capacity or permeability of soil depends on number of factors such as soil structure, texture, organic matter content, moisture content and slope percentage.

Avenues of Water Loss from Soil

Factors which influence behaviour of water in soil in the context of soil erosion have been discussed above. Here, it is useful to briefly highlight the avenues through which water disappears from the soil.

Infiltration

The downward entry or movement of water into the soil is called infiltration. When the water is introduced to the soil either by irrigation or rain, it penetrates the surface of the soil. The speed of movement or the depth covered in a unit time is termed as infiltration rate. It is usually measured in centimeters per hour. The infiltration rate is greatly influenced by the moisture content of the soil. The soil takes in or absorbs water rapidly, when water is just applied on the field surface. If water supply continues, surface becomes saturated and infiltration rate decreases until it becomes merely constant. The soil layer with the lowest transmission rate, either at the surface or below it, usually determines the infiltration rate. Surface sealing which is the formation of a thin compact layer on the ground surface rapidly reduces the rate of infiltration. The surface seal results from a breakdown in soil structure due to the beating action of raindrops and the action of the water flowing over the surface whereby, the fine particles are filtered around the larger particles to form a relatively impervious seal. Very sandy soils may have infiltration rates of over 25 cm per hour, whereas clay soils may have rates approaching zero. Poor soil structure, salinity and alkalinity reduce infiltration rate.

Run-off

Run-off may be defined as the portion of rain or irrigation water which leaves the field either as surface flow or subsurface flow. As long as the rate of rainfall (or irrigation water) reaching the soil surface is less than the infiltration capacity, all the water is absorbed into the soil. When the rate of rain- fall exceeds the rate of infiltration, shallow depressions are filled to overflow level, water begins to flow over the land towards streams. The water required to fill depressions prior to the beginning of surface run-off is called initial detention or 'depression storage'. Overland run-off usually reaches a tiny rivulet or channel which connects to the big rivers and ultimately to ocean. The time of travel from raindrop to stream channels depends on distance, slope, surface condition and depth of flow. Surface run-off is the run-off which travels over the ground surface to the channel of a watershed. The channels may range from rills to major streams. Sub-surface flow is that portion of infiltrated water which penetrates to a shallow depth, travels laterally and is intercepted by channels or emerges to the surface. Appreciable amount of sub-surface flow, relative to total run-off, is likely to occur where permeable surface soils lie over slowly permeable or impervious horizons at shallow depths.

Evaporation from the Soil

Evaporation is the process of vaporisation by which liquid water is transferred into the atmosphere. The rate of evaporation of water depends on atmospheric temperature, soil humidity, organic matter content and wind velocity. Dry air has a greater capacity for absorbing moisture than moist air, hence evaporation increases under conditions of low humidity. High evaporation occurs at places of high temperature, low atmospheric humidity and high winds.

Transpiration by Plants

Transpiration is the discharge of water as vapour into the atmosphere by the leaves and stems of living plants. The amount and frequency with which irrigation water must be applied will be governed in part by the rate at which crops transpire. This rate of transpiration varies from practically nothing at young seedling stage during cold cloudy weather to a maximum during hot windy weather at a time when crops are growing most luxuriantly. Rapidly growing crops of a leafy nature may use about 5 cm of water per week during the peak growing season. The volume of water transpired by the plants depends mainly on the water at their disposal and also on temperature and humidity of the air, wind, intensity and duration of sunlight, stage of development of the plant, species of the plant and nature of the leaves.

Deep Percolations

When the amount of water entering the soil becomes greater than its water holding capacity, deep percolation occurs. This is an intensified form of infiltration in which a great loss of water takes place. Water goes beyond the exploration capacity of plant roots. This water moves due to the force of gravity.

Water Conservation Methods and Practices

It is estimated that about 46 per cent of the water received through rainfall is lost through surface run-off. Now it is visualised that if the rainfall water is obstructed and stored in soil at the place where it is received (*in situ*) it can help to reduce not only soil erosion but siltation of river beds and to tanks and ultimately control floods in plain. In addition to ! this, the water stored in soil can support good crops and increase the production markedly.

Table. Land Capability Classification.

Sl. NO	Class of land	Slope %	Adopted land use and soil conservation meausres
A. Land Suitable for Cultivation			
1	Class I	0-1	Any crop can be grown with proper crop rotation and green manuring to maintain soil fertility
2	Class II	1-3	Contour farming, contour strip cropping and cover cropping contour-bunding.
3	Class III	3-5	Intensive agronomical measures such as contour cropping, contour strip cropping, terracing or contour bunding
4	Class IV	5-8	Contour bunds or terraces and intense agronomic measures. Mostly soil building and soil maintenance crops are to be sown.
B. Land not suitable for cultivation			
5	Class V	8-12	Permanent pasture with controlled grazing
6	Class VI	12-18	Pasture, grasses and forestry. Grazing should be restricted
7	Class VII	18-25	Forest with restricted felling. Contour trenching as conservation measure.
8	Class VIII	>25	Forest with complete closure to grazing and felling of trees.

The term conservation means preservation of something against loss or deterioration. Water can be well-preserved by utilising the lands appropriately and erecting some structures. The concept of using the lands according to their capacity in order to conserve soil and water is termed as 'land capability classification.

Using lands according to their capability, ensures use of land without its damage and sustains crop production on a pennanent basis.

Some of the very useful water and soil conservation techniques which can be adopted at the individual watershed level, according to their suitability to the local agro-climatic conditions are put under (A) agronomic measures, (B) agrostological measures, (C) mechanical or engineering measures, and (D) afforestation measures.

Agronomic Measures

Following agronomic measures can be adopted for con- serving water and soil.

Contour Farming

Contour farming is practiced in sloppy or hilly regions. In such areas, the rain water is absorbed in very little amount because of its quick downward movement on the slopes. If these areas are ploughed up and down the slope, the heavy rainfall may cause gully development. Taking into consideration this effect, the sloppy areas are ploughed and seeded across the slope *i.e.,* in circular furrows around the slope. This process is termed as contour farming. The contours (circular or spherical furrows) catch the downward moving water until it is absorbed in the soil. The ridges reduce the flow of water. The circular rows of the crop plants across the slope, reduce the speed of moving water. Greater amount of water is soaked by the soil. Reduction in amount and speed of run-off results in considerable reduction in soil and water loss.

Contour is an imaginary line which connects the points located on the land surface at the same heights. Before raising a furrow or ridge, contour line should be marked with the help of wooden sticks or small brick pieces/pebbles. Furrows should be opened parallel to this marked contour soil. These furrows or ridges should be made sloppy towards the main waterways. Tillage should then be done parallel to the contour line on both sides. On small fields of uniform slope, one guideline may be sufficient, on large and more regular fields, several lines may be required. On a hil1side, first contour should ordinarily be laid about 1 to 1.5 m vertically below the top of the hill. Additional contour lines are located with the same spacing as the first. In ploughing, the best plan is to start back furrows on the guidelines and work around them.

Tillage Operation

Tillage operation includes ploughing, discing, harrowing and cultivating. The effect of tillage on water conservation may be both useful and harmful. Tillage is useful when it increases infiltration rate of water into the soil. It is harmful when it destroys soil structure, seals soil surface consequently decreasing infiltration and increasing run-off.

The concept of minimum tillage can be advantageously applied for soil and water conservation. In this process the land surface is left least disturbed.

As a measure of tillage operation, surface mulches are also used to prevent water and soil against draining away. These mulches reduce evaporation, increase infiltration capacity of soil, keep down weeds, improve soil structure and virtually increase crop yields. Crop residues like the stubbles, leaves, twigs etc. are used for mulching.

Crop Rotation

Crop rotation involves incorporation of legumes with cereals in a sequence in order to take advantage of different feeding zones of soil, both for, nutrients and water and to off-set disadvantages of mono-cropping in controlling insect pests and diseases. In an experiment at Dehloadun it was found that maise - pea rotation gives a net return of `1622/ha as compared to maise — fallow `988/1111 or maise — wheat `I079/ ha. At another place, two year rotation of green manured bidi tobacco — kodo millet (I year) and pigeon pea mixture of Sundia sorghum (II year) under rainfed conditions has been found suitable for rainfed condition. Another rotation *i.e.* pearl millet mixed with green gram rotated with green manured bidi tobacco, is also found suitable for dryland regions.

Crop rotation helps in water conservation by improving soil structure. It improves soil aggregation which results in increased soil permeability and increased water retention.

Besides crop rotation, mixed cropping is also useful from the point of view of conserving moisture and soil. Advantages of this type of croppings are good crop cover, feeding of crops from different soil layers and under fainted condition ensures the farmers against a total crop failure. In medium black soils of Kota, cultivation of sorghum mixed with pigeon pea has been found useful in reducing run-off.

Strip Cropping

This system of cropping involves growing of a few erosion resisting and erosion permitting crops in alternate strips on contour or across the slope with the objective of breaking long slopes on farm and preventing water and soil loss.

Crops that can be grown with minimum stirring of soil are, in varying degrees, erosion resisting or soil conserving crops. These crops are often close-growing, and nutrient enriching. Most of the legume crops and grasses are erosion resisting, whereas the tall growing, widely spaced crops like maise, cotton, jowar, bajra etc. are erosion permitting crops.

Strip cropping is quite often used in conjunction with other water conservation measures like crop rotation, contour farming, cover cropping, terracing, bunding etc. The following are the four general types of strip cropping :

1. *Contour strip cropping :* In contour strip cropping, the crops are arranged in strips or bands laid on the contour line, just across the natural slope of

land. In one strip, erosion resisting crop is grown and in next strip erosion permitting crop is grown. Usually the strips are cropped in a definite sequence.

2. *Field strip cropping :* Field strip cropping consists of strips of uniform width running generally perpendicular to the direction of the erosive force. For the control of erosion by water, the strips run across the general slope, but do not necessarily curve to follow any contour line. This practice is recommended only in areas where the topography is too irregular or undulating to make strip cropping practical.
3. *Wind strip cropping :* It is a farming system in which the strips of crop are raised across the direction of wind.
4. *Buffer strip cropping :* In buffer strip cropping, some grasses or legume crops are laid-out between contour strips of crops in the regular rotations. They may also be used as correction strips where the slope is irregular. The buffer strips may be wide or narrow. They may be even or may have variable depths. They may be placed only on the steep, badly eroded parts of a slope or they may be at more or less regular interval on the slope. The buffer strips are used to give more protection against water erosion than is afforded by a solid planting of grain or inter- tilled crop.

Addition if Organic Matter

Organic matter can be added to the soil in the 101m of farmyard manure, compost and green manure. Organic manures not only supply plant food elements but also help in improving the soil physical condition. They improve the rate of water infiltration by bringing about desirable change in soil structure and its permeability.

Organic matter helps in retaining moisture in the soil and greater amount of run-off water infiltrates into the soil. This infiltrating water contributes to the underground reserves of water. These reserves can be tapped in times of need.

Agrostological Methods

The following are the important agrostological practices that check water erosion :

- *Lay farming :* This method consists of growing grasses in rotation with agricultural crops. This practice improves water holding capacity and fertility of soil, helps in binding the soil particles and reducing loss of water and soil.
- *Retiring the land :* This practice helps in generating grass cover on the soil. Under favourable climatic conditions grasing should also be restricted. Numerous researchers have shown that grasses have good soil binding capacity and they greatly help the soil to absorb a large amount of rain-water. In different places, doobgrass (*Cynodon dacrylon*), *Dacrylis glome*

rata, *Eragrostis amabitis* and *Eragrostis cerbula* have proved their worth in conserving soil and water.

- *Afforestation and reforestation* : Tree plantations help a great deal in conserving rain water. The stems of the growing plants create a hindrance for moving water thereby allowing greater infiltration. Percolating water helps further vegetation to come up in the soil. Roots of vegetation hold the soil and water. After their death and decay, empty cavities are left which accelerate water retention in the soil. A large amount of organic matter is contributed by the trees too, which helps in moisture conservation.

Extensive plantation strengthens ecology and the hydrogen cycle of the region. The chances of rainfall and consequent acceleration of further vegetative growth are greatly in- creased.

- *Restricted grasing* : Excessive grasing not only removes vegetation from the soil but also scratch and loosens the soil surface. In such a situation, rainwater finds easy way to run-off along with valuable upper soil. Since grasing cannot be completely avoided a system of restricted or rotational grasing must be adopted. In this system an area open to grasing for sometimes should be closed for the following years to facilitate regeneration of trees and grasses and restore a thick ground vegetation.

Mechanical Methods

Following are the mechanical or engineering methods adopted to conserve water and soil.

Contour Bunding

This practice of soil water conservation is widely practiced in Maharashtra, Gujarat, Tamil Nadu, Karnataka and Andhra Pradesh, Contour bunding consists of consulting an earthen embankment across the slope of the land following the contours as closely as possible. A series of such bunds divide the area into strips and act as barriers to the flow of water, thus reducing the amount and velocity of the run-off.

Cultivation is not done on the contour bunds, therefore, about 5 per cent of the land is lost under the bunds for cultivation, about 25 to 162 tonnes of soil and a large quantity of water can be saved annually by constructing contour bunds. In addition to controlling erosion and maintaining soil fertility, the construction of bunds helps in better infiltration of water into the soil, ultimately replenishing the underground water table. Experiences have shown that contour bunds constructed in shallow, medium (depth 45 cm) and medium deep soils have stood well and have been beneficial to farmers. Bunds in deep black soil have been a failure because of the nature of the soil which cracks during hot weather and cakes during the monsoon. This results in the bunds having series of cracks especially during first few years of their construction. Bunds can be constructed by manual labour or by animal or mechanically operated machines. Bulldozers and motor-grades are employed in large scale contour bunding works, When animal power is employed the buck scrapper is the most efficient implement. In case of manual handling, burrow pits are dug towards upper slope of the bund and the soil is laid on the contour lines. When two men operated mould-board plough is used, furrow slice is thrown towards the bund from both sides. The spacing between contour bunds depends on slope per cent and the specification on the depth of soil.

Table. Spacing of Contour Bunds.

Sl. No.	Slope of land (%)	Vertical Interval (VI) between Two Bunds (in Metres)	Horizontal Interval (HI) between Two Bunds (in Metres)
1	0 to 1	1.05	105
2	1 to 1.5	1.20	98
3	1.5 to 2	1.35	75
4	2 to 3	1.50	60
5	3 to 4	1.66	52

Table. Specification for Bund Cross-Section.

Sl. No.	Depth of Soil (Metre)	Base Width (Metre)	Top with Ratio	Side Slope	Height (Metre)	Area o Cross Sectior
1	Shallow soils (7.5 to 22.5 cm)	2.67	0.38	1 :5 :1	0.77	1.14
2	Medium soils (22.5 to 45 cm)	3.12	0.60	1 :5 :1	0.85	1.56
3	Medium deep (45-90)	4.25	0.60	2 :1	0.90	2.18

Graded Bonding

Graded bunds are extensively used in India with a view to making safe disposal of excess run-off in high rainfall areas. The base width of these bunds ranges from one to two metres. They are useful in the areas where soil is compact and less permeable. No cultivation is done on bunds. The designs of the graded bunds involves the fixing of the dimensions and the slope of the disposal channel so that the velocity of run-off flow can be kept under control. On steep slopes, a parabolic channel and an eat1hen embankment are erected. On less steeper slopes, providing the bunds along a suitable grade and allowing the water to flow behind the bund ensures adequately safe disposal of excess run-off water.

Terracing

Ten-ace is an embankment or ridge of earth constructed across a slope to control run off and minimise water and soil loss. A terrace reduces the length of the hill side slope thereby reducing sheet and the erosion of soil and water.

Studies on soil loss with water have shown that doubling the length at slope increases erosion by 1.4 times and shortening the length of slope through terracing greatly reduces soil and water loss.

Terracing is a costly measure and thus cannot be justified on crop lands that can be protected by less expensive measures. Agronomic measures such as contour tillage, crop rotation and strip cropping are enough in many sloppy areas. These measures alone may furnish enough protection where rainfall intensities are low. But where erodable soil, long slopes and high rainfall intensities prevail, agronomic measures may give only partial control. Terraces should always be supplemented with the best possible cropping practices, since they fail to capture water when used alone.

Following types of terraces are constructed to conserve water :

- *Bench terraces :* This method involves constructing wide step like platforms along the contour line on sloppy land. Bench terraces make possible the growing of cultivated crops in mountainous regions.

  Based on shape, bench ten-aces are of three types namely :

  - Level and table top,
  - Slopping outwards, and
  - Slopping inwards. The first one is suitable for the evenly distributed medium rainfall areas and deep and highly permeable soils. Slopping outwards type of ten-aces are effective and economical in areas with a permeable soil of medium depth; and slopping inwards types are useful in high rainfall areas.

- *Broad based ridge terrace :* In this practice, ridges or bunds of 25-50 cm height and 5-9 m width with gently slopping sides are constructed. On these bunds cultivation can be done. Under some conditions base width is kept narrow and the distance between the bunds is also less. Such kind of terraces are called narrow based terraces. These are similar to graded bunds.

- *Channel type and ridge type terraces :* The channel type terraces are mostly used on soils where controlled removal of water is of prime importance. The important feature of these terraces is to dig a channel towards the inner end of the platform. Ridge type terraces are usually used where the water holding capacity is the prime function. Since spreading of water over maximum area is the main concern, the ridge is an important item and the channel is only incidental to produce the ridge. The ridge type of terraces are usually built from both sides. The base width of a narrow based terrace is usually 1.2 to 2.5 m and is more for the broad based ones.

However, the ridges of narrow based terraces cannot be cultivated unlike broad based terraces.

Gully Plugs and Check Dams

While planning gully controls, the economic factor should be assessed well in advance. Some times gully control measures may cost more than the value of area where the gullies occur. However, in such cases gully control may be justified on the basis of protection to adjacent or downstream land, reservoirs, buildings or other structures.

The various methods employed for controlling run-off have some underlying steps :

- Retention of run-off on the drainage area or catchment area by employing methods like contour farming, contour bunding, strip cropping, terracing etc. Diversion of run-off from above the gullies.
- Conveying run-off safely through gullies.

Following gully structures which are known as check dams, gully plugs etc. can be erected to reduce amount and velocity of run-off through gullies :

- *Temporary checkdams :* They are constructed across the bed of gully and have two uses :
  - To collect enough soil and water to ensure growth of pro- tective vegetation; and
  - To check channel erosion until sufficient vegetation is es- tablished.

The life of a temporary check dam ranges from 3-8 years and during this period enough vegetation is expected to come up. The check dam should be placed in such a way that the top of one elevates the base of another located upslope side. This criterion is considered in deciding distance between two check dams. The height should be more than 75 cm. In heavy soils, a slope of 0.1 to 0.5 per cent is given. An overall protection is provided by placing wooden planks, loose rocks or constructing platform of sufficient length and width which catches water falling from over the top of the check dam. These are placed towards down stream side of the check dams. Thus three types of temporary check dams can be constructed.

These are described below :

- *Woven wire dams :* Woven wire dams may be used as temporary aids in the establishment of vegetation for the permanent control of erosion.

To construct the woven wire dam a row of posts 60 – 90 cm deep is set along the curve of the proposed dam at about 1.2 m interval. Heavy gauge woven wire laced over the posts at 25-30 cm apart stm1ing from 25-30 cm above the bed level. The stakes are then driven down on the bed 30-60 cm deep. Paddy straw, sugarcane tops etc. along with stone pieces and soil are then packed along the row of post. The centre of the dam should be low enough to provide the necessary spill capacity.

- *loose rock dams :* In the places where rocks and stones of appreciable size and suitable quality are available, the dams could be constructed using these. The stones are usually encased in woven wire. To construct a dam, a trench is first made in the gully bed across the width to a depth of about 30 cm. This forms the base of the dam on which the stones are laid in rows and are brought to the require height. The centre of the dam is kept lower to provide space for spillway.

- *Plank or slab dams :* In building these dams, the posts are set in a row across the gully to a depth of one meter and about 1.2 m apart. The posts should be so arranged that one of the interval between them will fall in the approximate centre of the gully to form a spillway. After the post is set, a narrow trench is dug along the upstream side. The trench is made about 30 cm deep and just wide enough to permit placing of the headwall. The required spillway notch is cut out after the dam is constructed or a suitable opelling can be left at the headwall during construction. An earthfill is next made against the upstream face of the headwall and brought up to the height of the spillway crest. The fill should have slope not steeper than 2 : 1 and should extend well on the gully banks.

Construction of Permanent Structures

Permanent structures to conserve water and soil are built of masonry, reinforced concrete or earth. They serve as efficient supplemental measures in soil and water conservation. These structures are necessary at places where other less costly vegetative measures do not work. Permanent Structures are necessary where a high degree of safety or permanence is required. the main functions of permanent structure are to halt the advance of the vertical over falls at gully heads, to stabilise grades so that a gully can be changed to a regular waterway and in case of drop inlet structures, to control downstream flow.

Three permanent structures are erected to reduce loss of water and soil :

1. Drop spillways
2. Chute spillways
3. Drop inlet spillways.

Constructing permanent structures requires skill and thorough understanding of engineering principles and techniques. They art : costly and require substantial investment in the beginning. A well-planned system of water conservation requires supportive permanent structure like the ones cited above. However, their construction should be avoided if other vegetative and simple field structures are adequate to provide enough protection to soil and water.

Land Leveling and Grading

Land leveling or grading is the process of preparing or modifying the land surface to a planned grade to provide a suitable surface. Leveling requires cutting of elevated surface of land and filling the soil in depressed spots in order to remove surface irregularities and unevenness to make a plain surface. It makes suitable field surface to control the flow of water to check erosion and provide better drainage.

Land leveling is governed by certain factors such as soil characteristics, topography, cropping pattern, method of irrigation, rainfall statistics of the region etc.

The time for leveling should be such that manual, animal or mechanical labour is free from other farm operations. It should be carried out at proper moisture level so as to reduce the energy consumption in leveling. When soil is very dry, the unit draft required is about 3 kg/sq. cm as compared to 1 - 1.5 kg/sq. cm at optimum moisture. Before transferring soil, it should be loosened with disc plough or disc harrow. Bull- dozers are found to be very economical for leveling extensive areas in comparison to other equipments.

Stabilisation of Bunds and Outlets with Grass -plantation

Bunds, terraces and other structures are created by manipulating the soil so as to reduce the flow of water. Outlets are the structures through which excess water ponding against a bund or terraces, is safely disposed off. As these structures are earthen, they are vulnerable to erosive action of rapidly moving water. Growing grasses over these structures make them highly stable and resistant to damage. Extensive network of the grass roots keep the bund soil tightly bound together.

Construction of Tanks

Storing water on the surface and artificially recharging ground water reservoir is highly imperative in rain fed, arid and semi arid areas. Constructions of irrigation tanks or percolation tanks provide for the cheapest, considerably effective, hence popular means of harnessing the rain water in minor irrigation and village water supplying schemes.

Selection of Site

Intensive care should be taken while selecting the site of tank construction. It depends upon certain climatic hydrogeological and geomorphological features. Tank construction must be supported by soil survey, measurement of permeability in the zones, water table, infiltration rates in the top soil, rainfall atmospheric humidity, temperature, topography, drainage pattern, etc. Percolation tanks are meant for recharging the unconfined and semi-confined aquifers in areas where the annual rainfall is less than 75 cm and where ground water is over exploited.

Percolation tanks are suitable for the areas where :

- The surface soil is sufficiently permeable;
- There are no layers in the zone above the water table of such low hydraulic conductivity causing perched ground water mounds that rise into the tank and restrict infiltration;

- The regional water table is sufficiently deep to keep the rise in water sufficiently below the bottom of the tank and below the root-zone in the command area but not so deep that large quantities of water are needed to wet the rootzone before any recoverable water reaches the watertable;
- The acquifer is unconfined and sufficiently transmissive to allow lateral movement of the recharged water without building up water table right into the root-zone; and
- Permeable formations exist in the down stream side of the tank without any barrier formations like dikes in between.

Field Approach

Following points should be taken into consideration while planning for tank construction :

- If aerial photographs are available one can prepare soil maps, geological maps, bring out' buried channels, know vegetation conditions, bring out structural features and drainage characteristics of larger areas within a short till reducing the quantum of field work.
- Total catchment area and the run-off available should be assessed by using Survey of India top<) sheets and by analysing rainfall data,
- Physiographic features by using toposheets and aerial maps should he traced if available.
- Soil map, soil analysis, determination of soil permeability if possible and geological map of the area including the catchment area with structural details that control occurrence and movement of ground water are determined.
- Detailed inventory is taken of existing wells to know the type of formation to prepare water table maps for at least one complete cycle and conduct pumping tests to determine specific yield etc., and estimate present withdrawals.
- Geo-electrical resistivity and seismic surveys to determine the depth of bedrock, thickness of soil zone etc."

This information is useful in determining to what depth tank is to be dug, to what depth the foundation of the dam across the stream sources should be laid and to what depth the cut-off trench beneath the bund should be dug for building the core wall.

- The surface and sub-surface soil profiles, geological features and bedrock depths in the tank area, command area and along the stream course have to be determined. In shallow bedrock zones check dams should be provided along the stream course to tap seepage along stream.

Irrigation tank differs from percolation tank in that the main purpose of irrigation tank is to preserve the water surface in order to make its immediate use.

Therefore, measures are adopted to reduce percolation or seepage of water from the irrigation tanks. The field approach in locating the site for irrigation tanks is the same as that of the percolation tanks. Irrigation tanks are located in zones where the soils are less permeable and the bedrock is very shallow. The foundation of the dam across the stream course and the cutoff wall should reach more than 98 per cent of the depth to the bedrock to : prevent seepage losses.

Seepage or percolation can be reduced by following means :

- Reducing the wetted surface area in relation to the quantity of water stored (*e.g.* circular design);
- Application of sodium salt to calcium rich clay soils;
- Soil compaction;
- Tank silt or clay soil with bhusa;
- Mud plaster;
- Bentonite application; and
- Soil cement mixture.

Water Harvesting Ponds

For collection of run-off water two types of ponds are used namely (a) embankment types of ponds are feasible in hilly, and undulating topography where, by constructing a small length of dam across a water. course maximum storage of water can be affected. Depending on site conditions, storage/ earth work ratios of 5 to 20 can be achieved. In flat areas, embankment type of ponds are not feasible, In such areas, excavated ponds are constructed advantageously. In case of these ponds the storage to earth work ratios are relatively much smaller generally around one.

Selection of suitable site is an important phase in the planning and construction of a farm pond, The site selected should ensure that the catchment is neither too small nor to, big, too small a catchment may not contribute sufficient water to the pond while too big a catchment may result in rapid siltation of the pond. Catchments well-protected with conservation measures (*i.e.* bunds and trenches) will contribute less silt to the ponds. Other aspects like willingness of the farmer, both for construction of the pond in his area and for sharing the water with other beneficiaries, availability of sufficient command area with minimum undulation and free from soil problems like salinity, alkaliluty and stoniness are to be considered in selection of pond site. Deciding a suitable size for the pond requires information on the catchment size, its run-off producing characteristics including type of soils, land use, topography, conservation measures adopted and their effective storage losses (seepage and evaporation) from the pond and control measures adopted, period of storage required, area commanded by the pond and type of crops covered with protective irrigation. If other uses like providing drinking water for livestock and domestic uses are also involved, then these are also to be considered for deciding the total amount of water required to be stored. Usually provision is to be made for meeting all the combined demands for 18 months so as to cover the deficit in case of drought. The size (capacity) of the farm pond is also decided upon the volume of run-off reaching the pond. The pond size is fixed to hold about 50 per cent of the run-off so that the pond gets filled even during lower rainfall years. Except for selected locations, the information on total volume of water expected at different places in India, is not yet available.

Embankment Ponds

In the embankment type of ponds, the total height of the embankment includes the design height of impoundment, a free board of 20 per cent of design height for protecting against overtopping and the occurrence of storms of higher frequency than the designed one and another 5-10 per cent extra height for settlement due to consolidation of the earth work. A sufficient top width is provided to serve as a road connecting both the banks of the water course. Upstream and down stream side slopes of embankment are usually in the ratio of 3 : 1 and 2 : 1 respectively. In case of embankment with porous soils, a core wall of impervious soil is provided to cut down the excessive seepage loss and also to keep the phreatic line in check. A key trench is provided to secure the embankment with the foundation to overcome sliding action.

Excavated Ponds

These ponds are usually constructed as square or rectangular ponds for easiness of construction. A convenient depth is chosen for the pond. This usually varies between 3 and 5 meters. Greater depths result in excavation costs disproportionate to the additional capacity provided. Greater depths; therefore, reduce the area occupied by the pond and hence the loss owing to evaporation, seepage rates become excessive due to the increased hydrostatic pressure. Excavated ponds are provided with controlled inlets of the chute spillway type so that gully is not formed at the inlet portion. Emergency spillways should be provided in the ponds to dispose off excess water. Measures should also be taken to reduce seepage from the ponds.

Use of Water Preservants

Water preservants are organic polymer compounds endowed with a great capacity to retain water. Their production and consumption is very limited in our country. However, they are very useful for small fields and dryland areas.

A recently discovered and the most efficient water preservant is known as *Jalshakti'*. *Jalshakti* is an off-white free- flowing powder which on absorbing water forms a smooth gel which acts like a water reservoir. It is a polymer, hygroscopic in nature and retains its free flowing character even on absorbing up to 50 per cent by weight of moisture. This granular organic compound is especially designed to improve plant soil relationship. *Jalshakti* is safe to use. It is non-toxic, bio-degradable and has no adverse effects either on the crop or on the soil.

Moisture pre servants are mostly used as soil additives. They are also sometimes used for seed treatment. On finding sufficient moisture in the soil they absorb many times more water than its own weight. They keep the water largely safe against percolation, seepage loss and evaporation when the water level in soil goes down in post monsoon season, moisture preservants act as water reservoirs and meet the moisture requirement of crops.

Chapter 25

SOIL CONSERVATION MEASURES

INTRODUCTION

Watershed management at present is one of the most accepted form of developmental approach. It begins with the soil and water conservation but is expected to move on to the natural and human resource management and finally to community building. Water is the most essential item for every living being. Soil is a porous material and acts like a sponge capable of absorbing and retaining the water. Thus soil acts as the vessel in which water is stored in the nature. This retained water is called soil moisture part of which is available to the plants. Therefore, the plants to obtain water from the soil there should be water stored in the soil fur this first of all the soil should be stored or preserved. The question of soil conservation becomes most crucial because soil is very easily eroded within no time. At the same time it should be remembered that it takes hundreds of years to form one centimetre thick layer of soil.

Most of the cultivated lands are slopping and require one or more types of soil conservation measures. The topography and agro-climatic conditions are so varying in our country that one should have the ability to discern what type of soil conservation measures a particular land area requires. This means he should have a clear idea of the various soil conservation measures and their suitability to a particular geographical area where the soil conservation is a problem. One should have the ability to diagnose the type of soil conservation problem and to propose appropriate treatment to solve each problem. Often he should apply a combination of treatments in which case he should have a clear idea of the combined effect of various in the soil conservation measures.

Most of the cultivated land need application of long-term soil conservation measures followed by re-organisation of land use pattern efficiently including sun moisture conservation, improvement of soil productivity and cropping and farming systems.

There are many ways of conserving the soil. Numerically speaking there are about 58 ways of mechanical soil conservation, 13 ways of biological soil conservation and 20 ways of agronomic soil conservation. Often both the trainers as well as trainees get confused between the various types soil conservation measures. Some of them of course are very much similar to each other. In the field : level situation a combination of a number of soil conservation measures have to be applied. Therefore one should have the ability to decide which combination of soil conservation measures should be implemented. This booklet intends to enumerate all the possible types of soil erosion control measures and their combined applicability to various conditions of soil degradation and erosion. Several of the measures enumerated may not be found in the conventional literatures. They are collections from the field practices observed by me in many parts of the country.

As already mentioned there are mainly three broad types of soil erosion control : mechanical, biological and agronomic. Under each of these there are many types of individually different measures. The readers are strongly reminded that it is a combination of several soil erosion control measures drawn from the mechanical, biological and agronomic types that becomes effective in the actual field conditions.

We cannot change significantly the climate and the inherent soil properties of a place. But we can do something to modify the other factors of topography and vegetation to reduce soil and water losses. However before we discuses each method we need to know the factors that should be taken into consideration in determining the type of soil conservation measures.

Factors to be Considered

The following factors and the mechanics of erosion by water and wind are to be assessed before we implement any of the soil conservation measures. They are briefly described here.

Physiography

The physiography refers to the size, shape, relief, drainage, mean elevation, land slope etc. Both the volume and rate of run-off increases as the size of the water shed incrcases. The size of the watershed is an important factor to compute the peak rate of run off which is essential for the designing of the drainage channel.

The longer and narrower the watershed the time of concentration of water (water retention in the water shed will be longer and lesser will be the run-off.

Relief designates the elevation difference between any reference point on the basin with respect to outlet elevation. Greater the elevation lesser will the concentration effect. At the same time longer the distance between the points greater will be the concentration effect or run-off water. Ultimately the combined effect of the elevation difference (slope factor) and the total length (length of slope) between the points will be the actual concentration effect.

Land slope has major implications for the land use. The speed and the extend of run-off depend on slope of the land. Greater the slope more will be the velocity of the water flow and still greater win the erosion of the soil. The complex mathematical expressions of the multiplying effect of the slope on the velocity, erosivity, the quantity of soil eroded and the particle size is expressed here in words as follows.

"When the slope is increased by four units the velocity of water flowing along the slope is doubled; if the velocity is doubled the energy and the consequent erosivity is increased four times; the quantity of soil eroded is increased by 32 times and the size of the particles that can be transported by pushing or rolling is increased by about 64 times."

When the slope is more the number of contour lines will be more per unit area. Based on this relationship we can estimate the percentage of slope of a given area using the following formula : $S = MN/A \times 100$, where S is percentage pf slope, M is total length of all contours within a watershed in metres, N is the contour interval and A is the area of the watershed (m^2).

The degree of slope determines the land use for annual crops, plantation crops, land reclamation, depending on soil depth, stoniness, etc. Hence the length and degree of slope is important parameters to be considered in the watershed management. Roughly we can say that up to 3 per cent of the area is good (or bunding, 4-7 per cent slopes is suitable of contour bunding, 8-10 per cent slope for broad base terrace, 11-15 per cent slopes for broad terrace, 16-20 per cent slope for broad bench terrace, 21-25 per cent slope for bench terrace, 26-30 per cent for narrow bench terrace, 31-50 per cent slope is suitable for horticultural and fodder trees and between 51-100 per cent slope the land should be used for planted forestry for fire wood and small timber trees and above 100 per cent the land should be left for natural forest. This division is arbitrary and is based on the practical experience in the field.

The total rainfall and the drainage density are factors that need to be considered for the watershed management. The factors affecting drainage are related to the susceptibility of soil to erosion, run-off pattern, sedimentation and locating erosion control structures and erosion itself.

$$\text{Drainage density} = \frac{\text{Total length of all streams (km)}}{\text{Catchment area}\left(\text{km}^2\right)}$$

The drainage map of an area may serve as a beneficial tool for the understanding and the preparation of the erosion assessment. From the drainage map we can understand the drainage pattern of an area referring to the design of the stream courses and their tributaries. The drainage may be fine, medium or coarse textured depending on the type of soil. Clay soil will exhibit finer drainage structure while the sandy soil will show coarse structure. Drainage patterns can act as guidelines to locate vulnerable areas requiring different kinds and degree of soil conservation measures.

Soil Properties

Soil properties refer to soil series and soil phases, physiccal, chemical and biological properties, hydraulic soil groups and soil moisture regimes. Based on the run-off potential there are four hydraulic soil groups :

1. Low run-off potential,
2. Moderately low run-off potential,
3. Moderately high run-off potential and
4. High run-off potential. The run-off very much depends on the infiltration and permeability of the soil.

Vegetative Cover

A dense vegetative cover is the most powerful means to reduce the soil erosion. In relation to the vegetative cover the following hydrologic conditions may be identified : (a) poor (heavily grazed or regularly burnt, litter, small trees and brush are destroyed), (b) fair (grazed but not burnt, there may be some litter but these are not protected), (c) good (protected from grasing, litter and shrubs cover the soils).

Land Use Practices

A record of the present land use practices followed by the farmers in a region is essential for further planning and re-organisation of land use according to its land use capability classification, to get sustained production.

Nature and Distribution of Rainfall

Rainfall data is collected at different locations in the watershed and the mean is estimated. Besides total annual average the duration, intensity, frequency etc. of the rainfall are also important information for the watershed management.

Prediction of Peak Run off Rate

The peak run-off refers to the maximum level of run-off happening during a period of very high storm which occurs once in several years. In soil and water conservation measures, the design of hydrologic structures, quantitative estimates of run-off rates, volumes and distribution are to be worked out. The channels and other structures are planned in such a way that they will be able to carry maximum run-off, which can be expected in a specified recurrence interval.

Floods and Droughts

Related to the peak run-off is the flood situation which also occur either every year or once in few years. Needless to say that the soil conservation measures implemented should be able to withstand the flood waters. Whereas in the case of drought the soil conservation structures should be such that they store sufficient moisture in the watershed area to tide over the drought period. Therefore we should have sufficient data on the flood and drought occurrence in the watershed area.

Socio-economic Factor

The watershed management begins with the implementation of soil conservation structures. But it should cross over to the economic programmes and finally it should culminate social development programmes. Even in implementing the economic and social programmes the cultural factors are to be taken into consideration.

Mechanical Measures

The mechanical measures for soil conservation consists of any structure that is erected to hold the soil and thereby prevent the erosion. They are described as follows.

Bunds

Bunds are ridges or embankments or long and narrow projections constructed on the surface on the land at selected places and in selected directions. They are constructed with a number of materials, in different. sizes, shapes and heights. For more details about the type and construction of bunds the reader is requested to refer the booklet No. 593 on "Bunds and Teuacing". In this booklet the types of bunds described are with reference to the construction material and way of construction. Bunds are also constructed in contour or graded contour line.

Mud Bond

The mud bunds are constructed using the soil. This is constructed when the slope is minimum. This is common in farm lands and in places where rainfall is average and above. The mud from the lower side of the proposed bund is dug up and placed in position to make the bund. For more details about bund making the readers may consult the Booklet No. 593 on "Bunds and Teuacing". In places where there is average and above average rainfall there should be drainage from the plots in order to drain out the water.

Loose Rock Bond

In loose rock bunds instead of soil, rock are used. The rocks are piled up in the form of a bund. Rock of different sizes are arranged in such a way that the small ones are fitting in between the big ones.

This is constructed in areas where the rainfall is less than average and the rainfall is erratic. Stone bunds are constructed in places where there are plenty of stones available at very cheap rates and to withstand the sudden on rush of rain water. They reduce considerably the velocity of the rushing water. However the water is not retained some soil may be retained.

Loose Rock Cum Soil Bund

In this type the space between the rocks is filled with the soil which acts as a cementing agent. Besides the above mentioned : advantages this type; is, able to retain the soil also.

Bund Cum Vegetation

Trees; shrubs or bushes are planted on the top of the bund; or on one side of it, By doing this the bunds are strengthened.

Walls

Walls of different heights are made in certain places for the conservation of soil and water. They are of different types.

One Sided Mud Wall

These are modified bunds constructed on land have more slopes than the places where bunds cannot be constructed nor terracing can be done immediately. The down side of the bund is beaten and pressed in such a way that it looks like a wall. The upper side though shows bunded structure merges with the land on the upper side. This is possible only in clay soil. Grass is allowed to grow on the side of the wall so that the wall will be durable.

Double Sided Mud Wall

In this type the wall is built higher and hence both the sides have to be built by beating and pressing the soil to make a compact wall. By growing suitable grasses we can maintain the wall durable. It is also highly advisable to cover the double sided wall with any material to protect it from erosion due to rain.

One Sided Stone Wall

Instead of mud stones are arranged at an angle corresponding to the angle of retention. This is much more durable than the mud walls.

Double Sided Stone Wall

Stones are arranged on both sides of the wall with moist soil pressed and compacted in between.

Loose Rock Wall

Rocks and stones are arranged without any soil of cementing material in the form of a short wall. This is useful to check and reduce the velocity of the rushing water and soil coming along with it under erratic rainfall conditions.

Terracesc

Terrace is the cut and levelled portion of the slope. Depending on the degree of slope the width of the terraces will vary. Greater the slope lesser will be the width of the terrace. Depending on the width of the terrace there are different types of terraces. With reference to slope the terraces may be bunt in three ways : levelled terraces, inward slopping terraces and outward slopping terraces. For irrigated crops like paddy levelled terraces may be better. When one needs drainage through channel then inward slopping terraces are better. But if the drainage is natural then outward slopping terrace is better. For more details about terracing the reader is advised to consult the Booklet No 593 on "Bunding and Terracing". In this booklet only the different kinds of terraces are enumerated.

Broad Base Terrace

Generally broad base terraces are made on land slopping between 8-10 per cent.

Broad Terrace

Broad terraces are constructed on lands slopping between 11-15 per cent.

Broad Bench Ten-ace

These are lesser in width and are constructed on lands with slopes ranging between 16-20 per cent.

Bench Ten-ace

These are still narrower terraces constructed on lands with slopes ranging between 21-25 per cent.

Narrow Bench Terrace

These are terraces of least width and constructed on slopes ranging between 26-30 per cent.

Continuous Ten-ace

When the terracing done along the whole length of the contour it is called continuous terraces.

Broken Terraces

When terracing is done in a broken not continuous form it is called broken terraces. They can also be called intermittent terraces.

Interlinked Terraces

In this the terraces are constructed in such a way that one terrace leads or linked to another terrace on either side in the upward and downward way.

Each terrace is shorter in width and length. The design looks like criss-cross arrangement of the terraced beds. The linkages between the terraces helps in moving bullocks and implements up and down from one terrace to the other on the slope. Also it forms a zig-zag path for the people to go up and down the slope.

Intennittent Square Terrace

These are square shaped platforms cut on the slope to plant trees or to conserve soil and water around the trees already planted. The trees may be fruit trees, plantation crops like coconut, rubber, cocoa, coffee, pepper, nutmeg, clove, aracanut etc. The trees may be planted in contour line.

Platform Terrace

When the intermittent square terraces made for a row of trees planted in a contour line and extended both sides the whole thing looks like a plat-form cut across the slope. This is very common for plantation crops planted in contour line.

Half Circled Terrace

Instead of square shaped intermittent terraces for the trees half circled terraces are cut on the slopes.

Gully Check

Gully is a common term used for a variety of soil erosion effects ranging from few centimetres to three metres in depth and up to 18 metres in width. In some places gullies are called channels or nallah in Hindi varying from small channel to very big ones. The gullies are classified into very small gullies, small gullies, medium gullies and deep and narrow gullies.

This classification according to me is too broad. For example very small gully according to the conventional classification varies from few centimetres to three hundred centimetres. The range is too much. Similarly in the other classification too the range of depth and to some extend the width is too much.

The gullies may be with or without perennial flow of water. The barriers created for checking the soil erosion through various types and sizes of gullies are commonly called gully plugging or gully checks or channel/nallaha checks/ bunding in general.

Gullies with water flow is called streams and the checks made in them is called check dams. While checks that are made in gullies with flow of water only during the rains (seasonal) are called gully checks or gully plugs. Since the flow of water in the gullys is so much forceful the checks should be strongly constructed. The needed strength of the checks depends on the force of flow of water (quantity of water, percentage of slope, length of the slope, type of the soil etc.) and duration of rain and flow. Therefore the strength of the checks should be varying by adopting suitable construction materials and method. The different types of checks described here are precisely according to their variation in their strength to withstand the flow of water. They can be constructed in a number of ways. Most of them are explained here.

The width of the checks obviously depends on the length of the checks. More the length greater will be the width of the checks. As a thumb rule we can say that for every 9-12 inches length the width of the check should be one inch (1 : 9- 12). Therefore for 10 feet long check 10 inches thickness is needed. However this will vary according to the place and according to the force of flow of water. The width of the checks given here does not include the width of the foundation. In some cases the design and nature of the foundation should be given very serious consideration. Otherwise the gully checks will not survive very long.

The gully checks should be permanently built. Only then it will achieve attain its purpose of preventing the increase in the depth and width of the gully and also level the gullies. Therefore it is highly recommended that only permanent. checks/structures should be built across the gullies.

In a gully there will be a number of checks installed at regular intervals (distance between two successive checks). The main principle to be followed under such conditions is to maintain proper interval between the checks. The top of the lower check should be at least one foot higher than the ground level of the upper check. Any thing less than this will result in the erosion of the soil. The height of the checks will be deter- mined according to the percentage of slope and the distance between two successive checks and the depth of the gully. For a fixed height of the check the interval will vary with variation in the percentage of slope : greater the percentage of slope lesser will be the interval between checks and *vice a vice*. Similarly for a fixed distance the height of the checks vary with the percentage of the slope of the gully. The length of the checks depends of the width of the gully. The type of the gully checks described here are in the order of decreasing strength.

Concrete Checks

When the force of the flow is very high strongly built concrete checks are necessary to control the gully erosion. Sufficient foundation should be laid at the bottom and on the sides. For the reinforced concrete checks normally 1 : 3 : 6 to 1. : 4 : 8 cement, sand and metal mixture is used for the construction of the concrete check.

Sised Stone and Cement

Well sised stones are used to construct the gully checks. Well constructed checks are as strong as the concrete ones. The cement mixture have the proportion of 1 : 4 cement and sand. Instead of sised stones we can use unsised stones also.

Brick and Cement

In this type well baked bricks and cement are used to construct the gully checks.

Loose Rock Double Sided Wall with Cement

Instead of the sised stones irregular stones are used for the construction of the checks. The stones are arranged in such a way that both sides of the wall will be in the same level. The stones are jointed with the mixture of 1 : 4 cement and sand.

Loose Rock Double Sided Wall with Pointed Cement

Checks are made of any irregular shaped stones collected from the vicinity and are arranged in such a way that there is a wall on both sides. The space in between may be filled with small stones or soil. Finally the space between the stones is filled with the cement

Loose Rock Double Sided Wall without Pointing

This type of checks are made in places where the force of flow is less compared to the previous cases and which allow the water to drain out through between the rocks while retaining the soil. Eventually the spaces between the walls will get filled up with the incoming mud and the wall becomes impervious.

Loose Rock One Sided Wall Pointed with Cement

The one sided walls are more suited to gullies with greater percentage of slope and in cases the upper side of the gully check is levelled at the time of construction. The construction of this type of wall is similar to the above mentioned one except that the stones are arranged in level on one side only. Obviously this type of wall is less strong than the previous one.

Loose Rock One Sided Wall without Pointing

The construction is same as the above mentioned ones except there will be no use of cement. On the upper side of this wall is filled with soil and stones.

Brush Wood Checks

This is a fencing structure across the gully and usually made in places where plenty of wood is easily available. The structure consists in vertically planted posts and the horrizontally fixed long poles. (Hereafter posts refer to vertical structures and the poles refers to the horrizontal, structures). The vertical and horrizontal structures may be made of any material such as concrete, iron, live wood and dead woods. The space between the vertical posts and the horrizontal poles can vary from very close to several feet apart. The space between the vertical posts and the horrizontal poles will be covered with twigs, leaves or any other vegetative material so that the soil eroded in the beginning will come and settle against them making a barrier to further soil erosion. In some case the space filled at the time of the construction itself. Whatever be the material of construction the brush wood check dams should be durable. When the brush wood check is made of live wooden material it can considered as a biological measure also. However it should be remembered that here we are discussing the mechanical structures.

There are several types of brushwood dams as explained here :

- *Live posts and poles :* The branches of trees that sprout in to new trees are planted as post. The horizontal poles are planted into the soil on both sides of the gully and tied to the posts. Both the posts and poles strike roots and sprout into branches forming an interwoven vegetative structure. Obviously this can be done, only in deep and narrow gullies having deep soil layer from the top to the bottom of the gully and in places where easily rooting and sprouting trees are available.
- *Live posts and dead poles :* The construction is same as explained before except that it can be constructed in places where the sides of the gullies are rocky or stony.
- *Dead posts and poles :* Dead wood collected from the nearby areas is used to construct the brush wood checks.
- *Concrete posts and poles :* Concrete posts and poles are either planted or constructed *in situ* to form the brush wood check.
- *Concrete posts and wooden poles :* The concrete posts are constructed *in situ* or brought from the place of fabrication. The wooden poles are fixed horizontally.
- *Live posts both sides of mod bond :* In this case a closely planted brushwood posts will be lining on both sides of a tapering and high mud bund constructed across the gully. In this case the posts planted are thinner than the posts mentioned above. They are branches of the trees which strikes root and sprout easily.
- *Dead posts both sides of mud bund :* Instead of the live posts dead wooden branches of any kind of trees are used.
- *Conically planted double line posts :* In this case the brush wood posts are planted in two Iines at suitable distance but slanting towards each other in pairs and crossing at the required height of the check. When constructed it will look like a conical fencing. Reapers are fixed horrizontally to strengthen the fencing. We can construct this type with live or dead posts including that of concrete and iron.
- *Straight and slanting line :* In this type of brush wood checks two lines of brush wood posts are planted at a suitable distance. The posts in the upper side line will be planted straight while the posts in the lower side line will be planted slanting towards the upper line and each post in the lower side line will be crossing with the corresponding post in the upper side line at a height required for the brush wood. Reapers may be fixed horrizontally to strengthen the brush wood structure. In this case also we can make it with live or dead posts including that of concrete and iron.

Sand Bags

Sand bags are easy material for constructing checks where there is water flow in the gullies. The sand bags are arranged in such a way that they form a well shaped check across the gully.

Building Sides

Often the sides of the gullies are prone to land slides or massive soil erosion causing the increase of the width of the gully. By building the sides one can retain the gully and used as drainage or a channel.

Paving the Bottom

The gullies could be paved with stones or bricks in case one wants to retain the gully and use it as a drainage or channel.

Gibbon Structures

The structure of gibbon constructions consists in having a very strong wire mesh holding small or big bolders in the form of a bundle. The shape of the bundle may vary from round to cylindrical. Gibbon structures can take the place of bunds, walls of terraces, sides of channels and gullies, checks, pillars etc., and are usually made where strong foundations are not possible to make the above mentioned structures. Galvanised wire mesh is used in the case of permanent gibbon structures whereas for temporary structures ropes and wires of lesser strength can be used. Based on the shape and material of construction we can identify several types of gibbons.

Gibbon structures can be used for gully checks, for one side walls to prevent land slides, to reinforce the side of the streams prone to land slides, to make double sided walls, stone path ways etc.

Galvanised Wire and Stones

Galvanised type of thick wire is used. Depending on the size of the stones the mesh size is decided. The mesh is spread on the ground in the place where the gibbon is to be constructed and the stones are piled up in the shape one wants. Then the mesh is taken over the pile and fixed firmly using proper wire tightening instruments so that all the stones in the pile will be held firmly in place. It will look like a bundle of stones but in any predetermined shape. We can construct gibbon structures in any size and shape and length. We can also construct them in layers one above the other in the case of permanent gibbon structures.

Ropes and Stones

Any type of rope material is used in the place of galvanised wire mesh. Though theoretically this is possible for practical purposes we cannot rely on this. Soil conservation is a long-term planned activity. Therefore gibbon structures with ordinary rope material will not be useful for soil conservation.

Gibbon Pillars

Wire mesh or rope mesh made in the form of cylindrical shape is filled with stones and finally when it is filled it will look like a pillar. Using these pillars further horizontal structures can be attached and made strong. Gibbon pillars can be used for constructing temporary check dams.

Mechanical Water Channel

The constant or frequent Row of water through the channels whether irrigation or drainage usually causes erosion of the soil in a massive way. The channels can be protected in several ways. The most common ones are explained here.

Concrete Channels

If the channels are big it is advisable to use reinforced concrete at the bottom and the sides.

Cementing

If the channels are small cementing the bottom and the sides would be sufficient for preventing the soil erosion.

Paving

Paving the bottom and the sides of the channel with stones, bricks, tiles etc. will help in controlling the soil erosion in irrigation and drainage channels.

Stone Spreading

Spreading bolders on the edges of the streams and water ways is another effective way of controlling the soil erosion.

Gravel Spreading

If the water channel or drainage is very gently sloping or almost levelled spreading pebbles or gravel would be enough to control soil erosion.

Trenches

Trenches are dug in different patterns across the slope to control the soil erosion by trapping the run off water. On lands more than 25 per cent slope trenches should not be made. In other words for a two feet wide trench the difference in height between the upper and lower edges should not be more than half a foot. Trenches are constructed in places where more rain water is received than that can be held by the bunds. Further the number of trenches dug in a unit area is more when the rain fall is high. Trenches of about one to two feet deep and wide are dug in any convenient direction or strength. For the control of the soil erosion trenches are made along the contour line. The following are some of the important types of trenches.

Continuous

When a continuous trench is made in contour line it is called a continuous trench.

Intennittent

These are trenches of shorter lengths dug in the countour or non-contour line with a certain amount of gaps in between. Where continuous trenches cannot be built due to rocks or stones intermittent trenches are dug.

Intennittent Slternate

When the intermittent trenches of two successive lines are alternated in position (gaps and trenches) they are said to be alternative.

Pits

Pits are either round or square. Pits can be dug on land (up to 33 per cent slope. Pits are more suitable where even intermittent trenches cannot be dug. This happens when the land is too stony or rocky. The pits are dug wherever possible.

Others

Besides the above mentioned soil conservation measures there are a few mechanical structure constructed for the control of soil.

Check Dam

Small dams built across the perennial streams to store or divert water and ultimately to control the soil erosion are check dams. Check dams may be built of mud, concrete, stones, bricks etc. For the details of construction of check dams the readers may refer to Booklets Nos. 573, 574, 579 & 576 on Investigation and planning, Design, Construction and Maintenance Earthern dams. In the case of check dam across the perennial streams the over flow is allowed by the side of the dam over a structure called spillway.

Anicuts

Anicuts are like check dams except that the over now is allowed over the dam itself. No spill way is constructed as in the case of check dam. Water is impounded on the upper side of the anicut.

Since the water is flowing and falling to the ground there is all possibility of cutting the soil at the foot of the anicut. In order to prevent this a check wall of about 2-4 feet called "shoe" is made a few feet away from the spot water falls at maximum flow. This is built of concrete or stones with cement in such a way that water falling from the anicut can be stored at a certain height of 2-4 feet. This water acts as an absorbent of the force of water falling.

Wiers

Wiers are structures constructed across the channels and streams. These structures have rectangular of triangular shaped notches to allow the water to over flow. Besides soil erosion control wiers are also used to measure the quantity of water flowing through the streams or channels.

Ridges and Furrows

Ridges and furrows are usually temporary mechanical structures though we can have them permanent too on a gently slopping land with planation crops, fodder crops, horticultural crops or forestry plantation. In seasonal cropping the ridges and furrows are temporary and small in size. In the permanent cultivation the ridges and furrows are made big. The effect of the ridges and furrows are suitable in low rainfall areas for the conservation water. According to the shape of furrows and ridges they may be classified into three types.

V-type

The ridges and furrows are made in such a way that between two ridges the furrow will be in the shape of English letter V. In this design the ridges and furrows will be small in size and the ridges will have sharp edge on the top.

U-type

The ridges and furrows are made in such a way that the space between two ridges will be in V-shape. The tope of the ridge will be round. The ridges will be bigger than the V-type. The bottom of the furrow will also be round.

L-type

In this type the furrows and ridges will be still bigger and the space between two ridges will be wider in the form of a channel with L-shaped sides. The top of the ridges may be flat with steep sides or broadly round. The furrows will be flat at the bottom forming like a broad channel.

Biological Measures

Biological agents like trees, shrubs, bushes, grasses are used in controlling the soil erosion. These are planted alone or along with mechanical measures. The trees, shrubs, bushes and grasses are planted in such a way that they are able to control the erosion to a great extend. They may be planted alone or in combination. However it should be remembered that the biological measures will have to be applied along with the mechanical measures in the areas where perennial cropping systems like forestry and plantation crops are cultivated. Whereas where seasonal cultivation is done one should; follow a combination of mechanical, biological and agronomic measures.

The various measures under biological control are so similar that one tends to over simplify them both in the understanding and practical application resulting in the failure of the programme. Each individual measure will be dealt here even though they are very much similar.

Tree Plantation

Land above 30 per cent slope should not be put under seasonal cultivation and the land above 45 per cent slope should always be put under perennial forestry which should never be subjected to clear felling. One can only go for selective felling of trees in areas above 45 per cent slope. But in some areas like origin of springs and streams should be kept under perennial natural forest at least 50 metres radius of the origin. From this area no felling should be done at all. Similarly areas which are so steep no felling should be done.

A word about selective felling is also necessary. Some of the ideas about selective felling will be too much for anyone to digest. However the ideas have to be projected before they are to considered. If we want to cut and fell a tree even selectively on to the ground some people have to go on the ground to the spot of the tree; it has to be cut manually or using mechanical saw; the tree will fall to the ground crushing some other trees, shrubs and bushes; then the branches have to be chopped and left in the forest if we want only the trunk; but if we want the branches also we have to cut them into manageable size. Then the trunk has to be dragged down to a place convenient enough to saw it into sawn timber or to load it into a truck. In some cases a road has to be constructed to the spot of the tree cut. In all these cases there will be unavoidable erosion of soil.

Therefore it is highly advisable that above 45 per cent slope selective felling of trees is done by means of helicopter. Helicopters should be used for people to slide down to the exact tree to be cut, chain the tree to the helicopter and then saw the trunk. As the trunk is fully cut and detached the helicopter should lift the tree along with the person who cut the tree to a convenient place, in the plains for chopping the branches and to saw the trunk into timber.

In this case all the branches, leaves, bark and waste from the sawing of the timber can be utilised. In other words every bit of the tree will be utilised. This method apparently looks very expensive. But if we calculate the cost of irreparable harms of soil erosion due road construction to the felling spot or dragging of the timber down through the slope and the cost of road construction or employing an elephant to drag the timber down to the plains and to load and transport the timber by truck and the cost of labourers etc., and the value of the amount of time required to complete the felling of one or few tree in the traditional way and considering the additional income from the branches and the leaves selective felling by means of helicopter will be much cheaper. From the point of soil erosion control alone helicopter felling is cheaper.

The point being driven here is that under no condition any land above 45 per cent slope should be left without tree cover even for a day. Because on that day a storm can occur and thousands of tonnes of soil can be eroded.

Besides this tree cover on land above 45 per cent slope is essential for the preservation of the springs and streams which are the sources of water for the big streams, small and big rivers.

It should be remembered also that the planting and caring of trees on land above 45 per cent is very difficult or almost impossible. Hence the best way perhaps is to saw the seeds already and allow the natural growth of the plants except areas having transport and accessibility. After having said the basic policy with regard to the utilisation of land above 45 per cent slope let us see the various methods of biological control of soil erosion. In all the tree plantation methods which are explained here the soil erosion is controlled by the canopy cover.

Random Plantation

Trees are planted at random with the same type of trees or different types. The first is called monoculture and the latter is called polycuture. Areal seeding also result in random plantation. Random plantation is done in areas where spring originates, both sides of the streams and rivers and in areas where human beings cannot reach.

- *Monoculture :* When the same species is planted in an area at random it is called random monoculture plantation.
- *Polyculture :* When different species are planted at random in an area it is called randomised polyculture.

Line Plantation

Trees are planted in line and there are different types of line plantation such as :

- Contour line,
- Square design,
- Rectangular,
- Hexagonal,
- Strip plantation,
  - Boarder plantation on the boundaries, edges of the streams, canals, roads, bunds, boundaries of plots etc. Line plantation can be done only in areas where people can easily reach.

Block Plantation

When trees of different species are planted in different blocks it is called block plantation. Each block may be of few to many acres. For commercial usages it is better to plant in blocks. Block plantation can also be done only in areas where people can reach easily.

Orchards

Planting of fruit trees is done according to the spacing required for each fruit tree. They may be planted in contour, square, rectangular or hexagonoal design.

Wind Break

Trees are planted as wind breaks across the direction of the wind. Wind breaks are planted on the edges of the seasonally cropped land and in orchards.

Shrubs Plantation

Shrubs are trees small in stature. They can be planted just like the methods used for tree plantation. Shrubs are more commonly seen in the dry and scanty rainfall areas. Shrubs are also planted along with the trees on contour bunds. Planting trees and shrubs together on contour lines at regular intervals will help the control of soil erosion considerably.

Grass Plantation

Grasses are excellent for soil erosion control. They can be planted on the bunds, on the edges of the terraces, on the side of the mud walls, in contour strips, on the boarders of the cultivated land, as pasture lands. Grasses are excellent for reinforcing the bunds; terraces and edges of the streams and canals.

Others

Other biological measures employed for the soil conservation are live brush wood check, pastures, live fencing.

Live Brush Wood

Fresh and green branches of trees that can be planted by cutting are used as the vertical posts of the brush wood checks. When planted in the soil they grow like any other tree and form the permanent vertical structure of brush wood check. The branches can be planted fresh or after striking the roots.

Pastures

Patures are areas where fodder grasses are grown to graze the animal. By maintaining pastures in a place we can control the soil erosion to a great extend.

Live Fencing

Trees, shrubs and bushes are planted on boundaries of the plot to form fencing. When these fences are across the slope they contribute to the soil erosion control. Even if it is not across the slope or along the slope it will help in checking the soil erosion to some extend.

Agronomic Measures

The soil conservation measures adopted in the cropping systems and practices are called agronomic measures. In other words these are agronomic practices made into soil conservation measures. There are a number of the agronomic measures of soil conservation. However not all practices are suitable or practical in all the places under all the agronomic practices.

However it should be remembered that agronomic practices adopted for the seasonal cropping will not be of much use for the control of soil erosion in the long-run. In other words it will be useful only for one season. After the crops are harvested there is always the chance of soil erosion if the soil erosion factors are prevalent. Therefore it is very important that for the long term soil conservation measures we should employ mechanical and biological measures strictly before we apply the agronomic measures in any place. It should be again remembered that all the agronomic practices are not suitable for all the places. In each place according to the cropping pattern and cropping system not only the mechanical and biological measures of soil conservations are to be adopted but also the agronomic measures are also to be adopted.

Contour Cropping

When rows of the crops are taken in contour line it is called contour cropping. This method is mainly adopted for the cultivation of seasonal crops at lesser slopes though contour cropping can be practiced for perennial crops in higher slope areas up to thirty per cent slope.

Contour Strip Cropping

Different crops are grown in contour strips on the slope. For example maise is cultivated in strip consisting of a number of lines across the slope then either on the above or below the strip of maise a strip of *arhar* crop is sown. Then any vegetable crop is sown in another contour strip. Thus we can take as many crops as we want. In this method a deep rooted crop is alternated with a shallow rooted one. Soil erosion can be controlled to a great extend by this method.

Mixed Cropping

A number of crops are planted at random mixed or in lines one crop after another. Theoretically speaking certain amount of soil erosion control may be there during the crop- ping period.

Multiple Cropping

Multiple cropping is a method of growing several crops in sequence on the same land during the same year. In other words in the same plot during the same year crops are grown during the *kharif* season followed by crops in the *rabi* season and *zaid* season. Because the land is always covered under I one crop or other the chances of soil erosion is less.

Multistory Multiple Cropping

Multistory cropping is a method in which mixed cropping of varying canopy heights are grown together. In this we can have double storey, triple storey and four storey cropping : Different crops of canopy height and width are adjusted within the same land to maximise the use of land area and the air volume area.

The principle in the multiple cropping is to plant the tallest crop in the maximum distance between which the next tall crop is planted at the middle.

The next tall (third in height) crop is planted in the middle of the first tallest crop and the second tallest crop. The next tall crop is planted in between the second and the third.

In the multiple cropping it should also be noted that out of the three or four crops at least two should be perennial and the rest seasonal.

The following examples will make the point clear :

- Coconl, it cocao -banana -fodder grass for costal areas
- Neem -guava -pomegranate -ground nut for dry areas
- Subabul -citrus -berseem for temperate areas
- Sapota -guava -banana -fodder grass for irrigated areas
- Sheesham -sapota -citrus -berseem for temperate areas
- Drum stick trees -guava -lime -pine apple
- J ackfruit - coconut -aracanut -clove -banana –pineapple for humid areas
- Aonla -custard apple -periwinkle- aloe vera for very dry areas
- Pear -orange -plum -straw burry for hilly areas
- Mulherry- Karonda -Banana -lucero for temperate areas
- Date palm -casuarina -bur -aloe vera for desert areas.

Organic Farming

Addition of organic manure increases the binding capacity of the soil particles providing some sort of erosion resistance.

Silvi-pasture

Silvi-pasture means cultivation of trees and grasses in the same place. Due to the tree cover and the grass cover the impact of the rain and other erosion agents on the soil will be minimal.

Ley Farming

In ley farming crops are grown in between two year duration of pasture cropping.

Alley Cropping

In alley farming crops are cultivated in between rows of trees which are planted at suitable distance to take seasonal crops.

Agroforestry

Agroforestry means growing trees on land where crops are grown. The difference between alley cropping agroforestry is that alley cropping the trees are planted in lines at definite distance. Whereas in the case of agroforestry the trees are planted at random and in gaps among the crops and in the edges and corners.

Mulching

Mulching is covering the soil surface with any biomass or sheets. By this the soil erosion is controlled in the cropping land.

Minimum Tillage

Stirring and cultivating the soil increases the soil erosion. Therefore, practice of minimum tillage is recommended "to reduce the soil erosion in the cropping land."

Cover Cropping

Cover crops are grown to cover the soil surface and thereby reduce the soil erosion and retain the moisture in cropping land.

Green Manuring

Incorporation of green leaves into the soil is called green manuring in general.

It can be carried out in the following ways :

- *In situ :* Any green manuring crop is grown on the same plot and when they are grown up they are ploughed and incorporated into the soil. Usually legumes are grown for in situ green manuring.
- *Applying green leaves :* Green leaves are brought from out side and applied on the land : The leaves may be of legumes or non-legumes.

Perennial Cropping

Growing perennial crops like fruit and plantation crops on lands which are prone to erosion will considerably reduce the soil erosion.

Dibling the Seeds

Placing seeds of trees, shrubs and bushes in holes made by a peg or a stick at regular intervals in a line is called dibling. In this method there is no need for ploughing. Similarly we can dibble seedlings also.

Planting Potted Plants

A number of crops (trees, pulses, vegetables, oil seed crops) can be grown in pots or polybags in nurseries. They can be planted in the field along with the mud balls

Spot Planting

Spot planting means sowing seeds or seedlings on spots at proper and recommended distance. Only the spot is prepared in the round or square form leaving the rest of the area untilled. Planting of gourds is done in this manner.

Growing on Pandals

Growing creeping, climbing and spreading crops on pandals helps in soil erosion control as the pandals form a barrier to the rains.

Alternating Deep Rooted with Shallow Rooted

Crops with deep roots are grown during rainy season during which time soil erosion occurs more easily and planting shallow rooted crop during the dry season.

Conclusion

Soil is the source of all our requirements in life. Erosion of soil is erosion of human life and finally destruction of forms of life from the face of the earth. Soil should be preserved by all means and preservation of soil should be every body's concern. However most people have very little idea about the various means of the preservation of the soil.

This booklet enumerates and describes briefly all possible types of soil conservation measures. Some of them are not found in any of the literatures as they have been compiled from the field experiences of various people including one's own. However it should be remembered that all these measures will have to be applied in combination and not in isolation.

Though mechanical erosion measures are costly without mechanical measures we cannot implement effective soil erosion control on the steep slopes, stream banks, gullies, in the rivers, streams, sides of roads and canals. However along, with mechanical measures we should always incorporate the biological and agronomic measures to improve effectiveness of the soil erosion measures adopted. Also biological and agronomic measures are income generation for the community.

Chapter 26

Tillage Systems in Soil Conservation

Conservation tillage systems cover a broad spectrum of farming methods primarily aimed at conserving and managing crop residue to reduce erosion. In the past, a residue cover of 30 per cent was considered adequate for erosion control and became the goal. Experience and data have shown, however, that truly effective erosion control relies on a situation- and site-specific plan developed by the individual.

Because 30 per cent was the former definition of conservation tillage, research data used it as a reference point or standard. Crop residue management is becoming the focus when looking at tillage systems for erosion control. Now 30 per cent is just a point along the continuum and may or may not be the objective for a given situation.

The tillage system of choice is the single largest factor in effective residue management. Each additional field operation incorporates more residue and increases erosion potential.

Selecting the most appropriate system for a particular soil and cropping situation requires matching the operations to :

- Crop sequence.
- Topography.
- Soil type.
- Drainage.
- Weather conditions.

Tillage systems such as ridge-till and no-till leave more residue and offer greater erosion control. No-till is the only system that consistently leaves at least 30 per cent cover following soybeans. For residue management, rotating tillage systems to coincide with crop rotations is an option. For example, a notill system following soybeans and a chisel or disk system following corn can provide adequate erosion control after soybeans and allows for some tillage in the less fragile and more abundant corn residue. Rotating tillage systems, however, slows the development of improved soil structure and may require investment in more equipment.

To minimise equipment needs, the selection of a tillage system should consider all crops in the crop rotation across all the soils of the farm. However, there may be some soils or some crops which may perform better with a different tillage system than the system selected for the rest of the farm. In these situations, a careful machinery analysis needs to be performed to see if the yield benefits would cover the additional fixed and variable costs of two or more tillage and planting systems. Sometimes, this analysis justifies the additional equipment, such as with no-till drilled soybeans and/or wheat combined with a mulch-till corn. Other times, it may be desirable to use only one tillage system, such as with ridge-planting with continuous row crops.

TILLAGE SYSTEM DESCRIPTIONS

Chisel Plow

The chisel plow produces a rough surface and generally leaves 50 per cent-70 per cent of the existing corn or grain sorghum residue on the surface depending on chisel point selection, shank spacing, operating speed and depth. Straight, narrow points, about 2" wide, leave the most residue. However, following crops that leave fragile residue even narrow points bury too much residue to provide adequate protective cover.

Where erosion is not a primary concern, 3" or 4" wide, twisted points invert more soil and bury more residue. In the western Corn Belt and the High Plains, wide sweeps are sometimes used on a chisel plow in wheat residue to undercut weeds and leave residue on the surface.

Typically, chiseling is performed in the fall and is followed by one or more secondary tillage operations in the spring. The fall operation cuts and incorporates some of the residue, making it more susceptible to decomposition and overwinter weathering than undisturbed residue. Partially decomposed residue is easily broken and covered by secondary tillage operations, negating much of the effect of having selected chisel points which leave more residue.

On many soils, a single pass in the spring with a disk, field cultivator or combination tillage implement provides limited pesticide and fertilizer incorporation on fall chiseled fields. A second tillage pass provides more complete incorporation, but can decrease residue and erosion control. Spring chiseling affords erosion control during the winter and allows extended grasing of stalks. However, soil moisture evaporation following spring tillage can result in yield reductions, particularly in lower rainfall areas. Spring chiseling may also produce clods that could require additional tillage operations to prepare a suitable seed-bed. Compaction problems may also arise on wet soils with multiple tillage trips. Larger implements may be required to provide timeliness with spring tillage operations, especially in wet springs.

A chisel plow may clog in extremely heavy or wet residue unless stalk shredding or light tillage precedes chiseling. This additional operation increases fuel and labour requirements. Several combination tillage implements have coulters or disks mounted in front of the chisel shanks which often eliminate the need for a prechiseling operation. The coulters or disks are operated just deep enough to cut the surface residue. This reduces the chance for residue clogging in the chisel area. The coulters will redistribute residue having the effect of increasing the percentage of ground covered. The shank spacing on these machines is usually 15″ compared to 12″ for a conventional chisel plow.

Disk and Field Cultivation

The tandem disk harrow is the most commonly used tillage implement in the Corn Belt. Typically, the disk harrow is followed by a field cultivator for final seed-bed preparation.

About 40 per cent-70 per cent of the residue generally remains on the surface after a single disking of corn, grain sorghum or wheat residue. Generally, disking corn or grain sorghum residue more than twice buries too much residue for effective erosion control. One disking is adequate if using a field cultivator for final seed-bed preparation. However, in fragile residue, even a single pass with any commonly used tillage implement does not leave enough residue for effective erosion control.

Fall disking saves time in the spring but the erosion potential from wind and rain increases, and snow entrapment decreases. The tilled residue is exposed to weathering, becoming more fragile and more easily destroyed with subsequent tillage operations. A spring disk system minimises erosion during the winter and is well suited to adequately drained and lighter textured soils.

A disk incorporates herbicides and other surface applied products. However, this incorporation occurs at the expense of soil structure. A common problem is disking when soils are too wet. Disking wet soils results in non-uniform incorporation, creates clods that require additional tillage operations and leaves a compacted soil layer below the depth of disking that can restrict root growth and reduce yields, especially in dry years. Disking wet soils to dry them out does dry the surface layer but creates considerable compaction below the depth of tillage.

A field cultivator accomplishes secondary tillage immediately preceding planting. The amount of residue covered by a field cultivator depends on the amount of time and weathering since the primary tillage operation. If field cultivation occurs two to three days after primary tillage, the field cultivator may not reduce the residue any further. In some circumstances, particularly with corn residue, field cultivation soon after primary tillage redistributes residue and may slightly improve per cent coverage. On the other hand, if time between primary tillage and field cultivation allows for some decomposition, a field cultivator significantly reduces residue.

One pass of a field cultivator or combination tillage implement is an alternative to a disk. The most common one pass tillage system is used in soybean residue. While a one pass system is successful in terms of yield and reduced costs, there is not enough soybean residue to effectively reduce soil losses on fields prone to erosion.

Stubble Mulch

The blade plow, or sweep plow, a common tillage implement in the High Plains, cuts weeds at the roots and leaves most of the residue anchored at the surface with minimum disturbance of the soil surface. Blade plowing is typically a summer fallow operation after small grain harvest. It kills weeds and loosens the surface. In moist soils, particularly those with higher clay contents, a blade plow may cause soil smearing below the blade, thereby limiting its use as a spring tillage implement.

The blade plow should be used on a hot day when the soil is dry for maximum weed kill. Small weeds are lifted and uprooted so they dry out. Deep operation often leaves roots in the soil and weeds re-root. Attaching a rod weeder or mulch treader behind a blade plow to bring roots to the surface improves weed control. Using these attachments, especially the mulch treader, results in greater residue loss and soil pulverisation.

Using either a blade plow or rod weeder at progressively shallower depths provides a firm seed-bed while controlling subsequent weed growth. These implements work beneath the soil surface and are typically used in drier climates; 30 per cent or more residue can usually be maintained. However, substituting either a disk or field cultivator for secondary tillage greatly decreases the probability of leaving adequate residue for erosion control.

Using herbicides for weed control and planting after one pass of the blade plow reduces or eliminates secondary tillage operations, but causes problems at planting because the soil is loose and fluffy, making it difficult to cut the loosened residue.

Ridge-till

In ridge-till, crops are planted into ridges formed during cultivation of the previous crop. Ridge cleaning devices push residue and surface weed seeds off the ridge either during planting or during a separate, preplant operation. A band application of herbicide behind the planter provides weed control in the row. Crop cultivation controls weeds between the rows and rebuilds the ridges for the following year.

Ridge-till reduces erosion by leaving the soil covered with residue until planting. After planting, 30 per cent-50 per cent residue may be left, but it is not uniformly distributed. Residue-covered areas between the rows alternate with residue-free strips in the row area. For erosion control, the ridges should be 3"-5" higher than the furrows after planting and that ridges be shaped to shed water to the furrow. For the most effective erosion control, orient ridges approximately on the contour.

Crop rotation influences the suitability of ridge systems. Ridges are maintained year-to-year with a cultivator, making ridge-till well suited to continuous row crops. Two cultivations are generally required : the first loosens soil and controls weeds, the second provides additional weed control and rebuilds the ridges. For ease of planting, the ridges should be rounded or flat topped, and 6"-8" tall after cultivation. Proper ridge shape and annual maintenance are keys to a successful ridge system. Be careful not to damage or destroy the ridges by wheel traffic, particularly during harvest.

Level or gently sloping fields, especially those with poorly drained soils, are well suited to ridge systems. The elevated ridges warm earlier in the spring. This warming, combined with drainage from the ridge, allows soil in the ridge to be drier at planting than untilled, unridged soil. A ridge system is an excellent choice for soils that are often too wet for early spring tillage, especially in the northern Corn Belt where the growing season is shorter.

Ridge systems complement furrow irrigation. Ditching, furrowing or hilling for irrigation provides suitable ridges for planting the following year. Chopping stalks or performing a very shallow, high speed tillage operation removes residue from the ridges and aids furrow irrigation, especially on soils with higher infiltration rates.

In a ridge-plant system, row cleaning devices on the planter move a small amount of soil, residue and weed seed off the ridge top. Ridge-cleaning attachments include sweeps, disk furrowers or horizontal disks. Except for possible fertilizer injection, no soil disturbance occurs prior to ridge planting.

In contrast, some ridge-till systems use limited tillage prior to planting. Tillage is generally very shallow, disturbing only the ridge tops. This tillage smooths peakshaped ridges to help keep the planter on the old rows. Depending on the tillage implement used, some control of emerged weeds and/or incorporation of herbicide in the row area is possible. However, tillage also incorporates some weed seed, rather than removing it from the row.

One modified ridge-till system uses a rotary tiller with planting units mounted behind it. To maintain the old ridge and avoid excessive power requirements, operate the rotary tiller only in the top 2"-3" of the ridge. Limit rotor tines to the row area to provide a strip till configuration. Other ridge-till systems use mulch treaders, rolling stalk choppers or flexible harrows.

Major reasons for using one of these implements include :

- Rounding or flattening the ridge.
- Removing some of the residue from the ridge.
- Killing emerged weeds.

Regardless of the tillage implement used, keep operating depth very shallow. Plant stubble in the old row should remain visible. Chemical incorporation is not a goal.

Fall Strip-till

While the ridges in the ridge-till system worked quite well to provide drainage on poorly drained soils, many producers prefer a strip tillage operation rather than ridges to aid soil warming. This zone tillage operation removes the residue from the row area, allowing sunlight to hit the soil surface and warm the soil. As with ridge planting, planting with strip-till takes place in the residue free strips but usually without the ridge. The adoption of no-till drilling soybeans in narrow rows made ridges unpopular in many areas.

Early strip tillage attachments were typically placed directly on the planter. One, two, or often three fluted coulters tilled and loosened the soil, burying some residue and working air into the soil. This aided in soil drying and warm-up but usually pulled up wet, sticky soil which interfered with the planter units. To reduce this problem and to allow soil warming to start sooner, some producers placed the coulters on a separate toolbar and performed the strip tillage operation anytime from one-half day to several weeks ahead of planting. Spiked-wheel residue movers and fertilizer knives have also been used to remove the residue and loosen the soil in these zone tillage systems.

While many variations of strip-till have evolved, the most widely practiced strip-till systems perform the tillage in the fall and usually place nutrients in the tilled zone at the same time. A sub-soiler, fertilizer knife, or multiple coulters perform the tillage allowing fertilizers to be injected through the knife or behind the coulters. Often, some type of soil mounding disks, such as anhydrous knife slot closing disks, build up a slight ridge in the tilled zone to aid in soil drainage. In drilled soybean residue, producers equip their strip tillage implement with row markers to properly lay out the cleanly tilled strips for planting next spring.

Some do not consider fall strip-till as a separate tillage system but as a variation of either the no-till or ridge-till systems. The Conservation Technology Information Center's definition of no-till includes strip tillage, provided less than one-third of the total row area is tilled. Some no-tillers may perform the strip tillage operation in the spring. Or as another variation, some ridge-tillers use a strip tillage operation on their ridges to inject nutrients and aid in soil warming.

No-till

Tillage is essentially eliminated with a notill system. Crop seed is placed in a narrow strip opened with a coulter or disk seed furrow opener. With little or no modification, most planters can be used in notill systems.

Common attachments include coulters, stronger down pressure springs and extra weight for better penetration. By disturbing only a narrow slot in the residue covered soil, excellent erosion control is achieved. Compared to other tillage systems, notill also minimises fuel and labour requirements.

Do not shred standing residue prior to planting. Planters, drills and cultivators perform better when residue is standing and attached to the soil, rather than unattached and lying flat.

Notill requires surface application of pre-emergence or post emergence herbicides for weed control. One or two properly timed applications may be necessary. For later planted crops (*e.g.* grain, sorghum, soybeans) apply an early preplant herbicide at 1/2 or 2/3 rate early in the spring when the probability of rainfall is high to provide maximum herbicide performance. This application often eliminates the need for a burndown herbicide and because weeds are controlled early in the growing season, soil moisture losses are reduced. For full season weed control, use a second lighter pre-emergence herbicide application at planting, a post emergence treatment, or crop cultivation. Equip cultivators to handle residue without clogging.

Notill planting is well suited to many soils. Residue, when uniformly spread, increases water infiltration and reduces soil moisture evaporation.

Using notill in poorly drained soils covered with large amounts of residue, delays soil warming and drying in the early spring, which delays germination and emergence. When colder, wetter soils are a concern with early planting dates, use notill planter attachments designed to move residue but not soil away from the row. Fall strip till facilitates warming and drying of the row area and may be a better choice.

IMPACT OF TILLAGE FARMING

Soil loss on arable agricultural land is typically an order of magnitude higher than under undisturbed native vegetation, and two orders of magnitude higher than rates of soil formation. There is much evidence to show that soil erosion due to rainfall and overland flow is exacerbated by tillage operations. However, of similar importance is the extent of tillage erosion resulting directly from tillage operations. This generally results in a movement of soil from convex shaped to concave shaped landscapes and leads to a nutrient-rich soil in the latter. While water erosion is strongly controlled by soil characteristics such as soil stone level, texture and crusting potential, experimental studies have shown that tillage speed, depth, direction and implement characteristics are the primary controls on tillage erosion. It is of major importance that eroded nutrients and sediment are retained in-field so as not to impact on surface water quality.

Given that rates of soil redistribution in the medium-term are influenced by tillage displacement as well as water erosion, it is necessary to separate these two components of soil redistribution in order to obtain a reliable assessment of water erosion rates. By using a tillage erosion diffusion-type model based on the one Lobb *et. al.* proposed and land use databases, Van Oost *et. al.* estimated that the mean gross tillage erosion rates for the part of Europe covered by the CORINE land use database was 3.3 Mg ha^{-1} yr^{-1}. For the same land area, they estimated the average water erosion rate was 3.9 Mg ha^{-1} yr^{-1} by using water erosion estimates for arable land, orchards and vineyards compiled in a study by Cerdan *et. al.* of datasets from 81 experimental sites across 19 European countries. The model also used large-scale land use (CORINE), soil (Soil Geographical Database of Europe), topography (Shuttle Radar Topography Mission) and soil erodibility datasets for Europe. For the cropland area of Ireland, the same models estimated the average tillage and water erosion rates to be 2.9 and 4.4 Mg ha^{-1} yr^{-1}, respectively. These erosion rates are higher than average rates of soil formation (consisting of mineral weathering, soil biomass growth and dust deposition) which range from 0.3 - 1.4 Mg ha^{-1} yr^{-1}, with the lower limit being indicative of European conditions. Research on tillage soils in Ireland is needed to validate the estimates of erosion rates given by the model of Van Oost *et. al.*.

The potential TP losses associated with these estimates of erosion could have serious implications for water quality in Ireland if the eroded sediment reaches surface waterbodies, given that the typical range for TP content of non-polluted agricultural soils in Ireland is estimated at between 0.02 to 0.2 per cent with the median TP content of Irish soils (0-10cm) being 0.11 per cent. Applying this range of TP in Irish agricultural soils to the estimates of water erosion reported by Van Oost *et. al.* for Ireland gives a range of 0.88 to 8.8 kg P ha^{-1} yr^{-1} lost from arable land. This is a conservative estimate of P loss from arable fields due to erosion, given that P has low solubility and is primarily bound to finer soil fractions like clay, which run-off preferentially transports. In a study of TP export coefficients from different CORINE land cover classes in 50 experimental sub-catchments of the rivers Colebrooke and Upper Bann in Northern Ireland, Smith *et. al.* determined the TP export coefficient from non-irrigated arable land to surface waters to be 4.88±1.12 kg P ha^{-1} yr^{-1} with 95 per cent confidence limits. This export coefficient was almost twice as high as that measured during the study for any other CORINE land cover class and almost five times as high as the export coefficient for pasture.

With the exception of tillage erosion occurring adjacent to waterways, soil transported in the field by mechanical tillage operations is unlikely to reach surface waters without transportation by water erosion. Though tillage erosion does not have the same direct detrimental effect on surface water quality as water erosion, it can increase the risk of nutrient delivery to waterways by progressive accumulation of nutrient-rich sediment in low-lying areas of fields which may be exposed to concentrated overland flow and leaching, and, therefore, it must be accounted for in any assessment of soil erosion. If field- and catchment-scale research is to identify sediment sources and test mitigation options within arable areas, it must be designed to explicitly attribute losses to tillage or water erosion processes. This will require assessment of each type of erosion in isolation and while interacting with one another. This is essential if we are to understand the role played by tillage erosion in delivering sediment to surface waters.

Rainfall variability in Ireland often results in tillage field operations being carried out in less favourable conditions (soils at or near field capacity) with increased risk of soil compaction from field machinery traffic. The trend towards larger machines with increased axle loads further increase the risk of soil compaction. Compacted soils with poor structure are more prone to surface capping and poor infiltration of water due to reduced porosity and consequent reduction in hydraulic conductivity, leading to earlier saturation and thus increased surface run-off and erosion in sloping areas. Soil compaction can occur as surface compaction *i.e.* within the tilled layer or as sub-soil compaction which occurs beneath the plough layer. Surface compaction is normally dealt with in the next tillage operation while sub-soil compaction is much more persistent and difficult to remove. While sub-soiling has been the subject of much research and can reduce bulk density and compacted layers, it is generally considered better to avoid sub-soil compaction than to rely on alleviating the compacted soil layer afterwards. Prevention of sub-soil compaction is essential for economically and environmentally sustainable agriculture.

Compaction can be reduced by :

- Use of low ground pressure wheel equipment on machinery;
- Working in good soil moisture conditions and minimising the weight of machinery;
- Minimising the number of passes of machinery and
- controlled traffic systems.

Measuring and Quantifying Soil Erosion on Arable Land

There is a need for information on both gross and net erosion rates from agricultural land, so that the sediment delivery ratio, or proportion of the sediment mobilised by soil erosion that is transported towards local water courses, rather than being deposited close to the original source, can be determined. If the level of erosion of Irish tillage soils is to be accurately determined, work must be undertaken that quantifies rates of soil movement to surface water at the catchment scale.

Traditional monitoring techniques used to establish soil erosion rates have the inherent flaw of failing to determine the fate of eroded sediment and, therefore, give no indication of the impact of measured erosion rates on surface water quality. Blake *et. al.* note that it is particularly difficult to assemble information on the spatial distribution of erosion and deposition rates within the landscape and on the associated sediment delivery ratios using traditional monitoring techniques. Much of the information available on erosion rates has been collected from flume and erosion plot studies; however, these only provide information on the net rate of soil loss from the bounded area, as represented by the flux of sediment across its lower boundary. As such, plot studies typically overestimate erosion rates by failing to encompass major catchment sediment stores. These stores get larger as catchment area increases because the fraction of less steep slopes, like valley bottoms where sediment deposition occurs, also increase. It is for these reasons that the representativeness of plot results in terms of the wider landscape is often questioned. As the scale at which erosion is being studied increases from flume-to-plot and up to field- and catchment-scale, the parameters influencing this erosion change and, therefore, so must methods used to measure erosion. The use of sediment fingerprinting and composite fingerprints to determine the provenance of eroded sediment is one preferable method at larger scales which will be discussed later.

Measuring and Quantifying Phosphorus Transfer from Arable Soil to Water

Soil cultivation is a major factor contributing to an increased risk of particulate phosphorus (PP) transfer to water, but when reduced cultivation such as non-plough tillage is practised to decrease losses of PP, there can be a build up of P near the soil surface which increases the risk of DRP loss in surface run-off. However, P loss can also occur through drainage, with the most significant instances of downward movement of P through the soil profile being associated with excessive application of P in manure and fertilizer, in particular, on soils with low P-retention properties and/or significant preferential flow pathways (*e.g.*, cracking clay soils).

Phosphorus Mobilisation

Mobilisation is the first key step in the separation of P molecules from their source and includes chemical, biological and physical processes. These processes group into either solubilisation or detachment mechanisms, defined by the physical size of the P compounds that are mobilised. The detachment and transfer of non-dissolved P in association with soil particles is more pronounced where farming practices generate erosion and provides a physical mechanism for mobilising P from soil into surface waters.

The size threshold most commonly used to operationally define detachment is >0.45 μm and has been used for the threshold between dissolved DP and PP. Haygarth and Jarvis have argued for the inclusion of a third mode by which P can be mobilised for transport to water — incidental transfer of DP and PP occurring when fertilizer or manure applications, which are not incorporated into the soil, are coincident with onset of rainfall. They conclude that even though incidental transfer will include mobilisation and detachment, it should be kept separate from these mechanisms due to the unique circumstances leading up to its occurrence and control. The relative proportions of PP and DP in surface run-off is dependent on the complex interaction between climate, topography, soil type, soil P content, type of farming system and farm management.

Particulate phosphorus includes all primary and secondary mineral P forms, plus organic P, P sorbed by minerals and organic particles eroded during run-off. It constitutes the major proportion of P transported from cultivated land (75-90 per cent). Fang *et. al.* reported that PP contributed from 59 - 98 per cent of total run-off P for un-vegetated packed flumes.

Unlike most DRP, which is readily available for plant uptake, PP acts as a long-term source of P for sub-merged aquatic vegetation and algal growth, particularly in lakes where inflowing rivers deposit nutrient-enriched sediment on the lake floor. Phosphorus release at the sediment-water interface may occur in the following conditions : (1) during periods of anoxia or hypoxia; (2) by wind-induced resuspension and bioturbation; or (3) an increase in pH of the interstitial water.

The expensive nature of field experiments and inherent variability in natural rainfall has made rainfall simulators and laboratory micro-cosms a widely used tool in P transport research. Due to the complexity of soil erosion by water, field experimentation can be complimented by work in controlled laboratory experiments. While there are still some reservations regarding the use of simulated rainfall in place of natural rainfall, there is widespread support for the use of rainfall simulation experiments to obtain some estimate of the magnitude of potential losses from different land management systems, soil types, and landscapes.

Numerous studies — outside of Ireland — have utilised rainfall simulation to evaluate nutrient losses in run-off from tillage systems. Studies have also been conducted using laboratory rainfall simulation on flumes packed with tillage soil to predict run-off of SS and PP using simple soil tests and examine variability in mobilisation and transport of nutrients and sediment by overland flow across a range of soils. In addition, flume studies using concentrated overland flow as opposed to simulated rainfall have been used by Knappen *et. al.* to show that the effect of conservation tillage on soil detachment rates is a result of soil property modifications affecting soil erodibility, rather than a result of the surface residue decreasing flow erosivity. Laboratory-scale work such as this is essential in understanding erosion processes and in selecting suitable erosion prevention measures for further testing at larger scales.

Phosphorus Source Areas

The loss of P tends to be highly sporadic in nature and is often restricted to small geographic areas. In many regions, small portions of saturated land, known as variable source areas (VSAs), generate the majority of overland flow, the occurrence of which is largely independent of rainfall intensity. This type of run-off is known as saturation excess run-off. A VSA can contract and expand both seasonally and during storms as a function of precipitation, topography, soil type, soil moisture status, and water table level. The occurrence of high STP within a VSA results in a critical source area (CSA) of P.

A large proportion (up to 90 per cent) of P exported from catchments on an annual basis is generated from a relatively small portion of the catchment and during only one or two storm events. Tunney *et. al.* showed that 40 per cent of the total amount of DRP lost in run-off for 1997 from four grassland fields ranging in size from 0.5 - 14.5 ha was lost when about 150 mm of rain fell in a four-day period.

In contrast, a study of nutrient and sediment loss to water from agricultural grassland catchments of the Dripsey River, Co. Cork in 2002, found that more than 80 per cent of TP loss was for the five months of October to February, with a large proportion coming from about 10 storm events where high P concentrations occurred simultaneously with high stream flows. This evidence suggests that, while extreme rainfall events with large return periods like that reported by Tunney *et. al.* can be responsible for a large proportion of DRP lost over an atypical year, more normally one would expect P loss to be spread across a number of large storms throughout the year. In addition, research at plot-scale on arable land in the UK by Quinton *et. al.* showed that more frequently occurring smaller events accounted for a greater proportion of the P lost over a 6-years period than infrequent large events. It is important to note that losses in the study of Quinton *et. al.* were measured at the end of an erosion plot and that even though a smaller proportion of P was lost in larger events, these events have greater transport potential and are more likely to deliver eroded sediment and P to surface waters.

The identification of CSAs, where the potential for pollution is higher, has significant implications for RBMP, because the blanket application of a specific mitigation measure across an entire catchment will not be as cost-effective as its deployment solely in those areas where it is most appropriate. Pionke *et. al.* suggested that effective mitigation of P losses from agriculture must focus on defining, targeting, and remediating CSAs of P loss.

Hughes *et. al.* used field and catchment-scale ranking schemes to identify CSAs for P loss in Ireland. Outside of the VSAs, run-off may be triggered when the infiltration capacity of the surface soil is exceeded, usually following high-intensity storm events. Both saturation and infiltration-excess run-off occur in Ireland and, though the latter is less common, research has shown that it does occur. At field-scale, run-off collection is complicated in areas with a perched watertable. The divisions between saturated and infiltration run-off become difficult as a VSA is generated.

TILLAGE ORGANISATION COMPARISONS

Typical advantages and disadvantages of tillage systems are shown in Table. The most important advantage of conservation tillage is significantly less soil erosion. Fuel and labour requirements are also reduced with conservation tillage. This information is useful in determining the suitability of tillage systems or combinations of systems for various situations. However, base selection on specific soils and cropping circumstances as well as individual management ability.

Of the systems compared in the tables, the moldboard plow system has the greatest fuel and labour requirements for tilling and planting corn and soybeans. Compared to the commonly used disk system, notill saves about 1-1/2 gal/ac in fuel and 20 minutes of labour/ac.

Labour savings allow a larger area to be farmed without additional equipment or help. Even if increased acreage is not anticipated, more timely planting may result in greater yields. In addition, costs for tractors, tillage equipment and maintenance will be less with fewer tillage operations.

Regardless of the tillage system selected, spread residue uniformly behind the combine using either a straw chopper or straw spreader. A chaff spreader may be desirable, especially when harvesting small grains or soybeans with larger combines. Uniform distribution of residue and chaff reduces equipment clogging and provides more uniform soil conditions for planting, easier weed control and better erosion control.

Recent advancements in herbicides make weed control with notill easier than it used to be. Early preplant applications, longer-lasting residual herbicides, and a wide variety of post-emerge products are helping assure weed control success with notill.

Conservation tillage systems represent alternatives at a time when economics require flexibility in crop-production. The growing concerns about agricultural sustainability and the environment require reduction of soil erosion. Conservation tillage effectively and economically reduces soil erosion and the resulting sedimentation, a major water pollutant.

Carefully evaluate the need for each tillage operation and pesticide application. Systems with more than two tillage operations prior to planting need careful examination. Additional operations are often unnecessary and only increase soil losses, compaction and production costs.

MITIGATION MEASURES TO AVERT EROSION ON TILLAGE SOILS

Various land management practices have been shown to minimise erosion risk on susceptible soils : low erosion risk crops and cover crops, tillage timing and intensity and the use of buffer strips. For example, intensively cultivated soils amended with spent mushroom compost, a bi-product of the mushroom growing industry in Ireland, exhibited improved structural stability as measured by an aggregate stability test.

The UK Department for Environment, Food and Rural Affairs (Defra) highlights potatoes, winter cereals, sugar beet, maise and grazed fodder crops as having the highest erosion risk based on crop cover. To minimise erosion risk on susceptible soils, low risk crops like oilseed rape, which establish a crop cover earlier, should be sown. Furthermore, winter barley may be more beneficial than spring barley as it provides winter cover. However, wet weather trafficking may offset benefits.

Minimum (or minimal) tillage, which involves shallow cultivation to a maximum depth of 10 cm using a tine cultivator, helps conserve SOM, promotes aggregate stability and thus reduces erosion.

In Ireland, minimum tillage normally involves :

- Shallow cultivation using a tine cultivator or disc harrow to a depth of 75-100 mm immediately followed by rolling;
- Spraying with herbicide a few days prior to sowing, following a stale seedbed period of a number of weeks (where possible) to eliminate volunteers and established weeds; and

- Sowing with a cultivator drill to a target depth of 40 mm.

To date, the effectiveness of minimum tillage to reduce erosion has not been investigated in Ireland. Research in the UK by Deasy *et. al.* found that for 5 site years, trialled losses of SS and TP decreased by an average of 151 kg SS ha^{-1} and 0.3 kg TP ha^{-1} under minimum tillage compared to traditional plough cultivation. Contour grass strips have received some research attention and have been shown to reduce sediment losses by reducing slope length and by acting as a barrier to slow down overland flow. Deasy *et. al.* found that although minimum tillage, crop residue incorporation, contour cultivation and beetle banks (raised vegetative barriers placed on the contour) all have potential to be cost effective mitigation options for SS and TP losses, tramline management (disruption of the compacted tramline surface to a depth of 60 mm with a tine) is one of the most promising treatments for mitigating diffuse pollution losses as it was able to reduce sediment and TP losses by 72-99 per cent in four out of five site years trialled. As a management practice to reduce P loss from tillage soils in Ireland, Carton *et. al.* advised that attention be paid to tramline compaction and that if soils become severely compacted, corrective action, such as sub-soiling, should be taken where appropriate.

The Nitrates Directive (91/676/EEC), as implemented in Ireland, sets out crop cover requirements where arable land is ploughed between 1st July and 30th November. The regulations require that the owner/occupier take appropriate measures to provide for emergence of green cover from a sown crop within 6 weeks of ploughing. In the UK, as part of the cross compliance regime, farmers are further required to carry out a field erosion risk assessment as a means of reducing risk to acceptable levels. The validity of this approach to erosion risk identification was verified by Boardman *et. al.* Conservation tillage in autumn may reduce losses of soil and PP by improving soil structure. In Norway, ploughing and shallow cultivation of sloping fields in spring instead of ploughing in autumn have been shown to reduce particle transport by up to 89 per cent on highly erodible soils. Rational land use policies such as the promotion of 'set-aside' on erodible soils, use of grass strips on erodible arable slopes, and buffer strips in riparian zones were identified as mitigation options to reduce soil erosion by Fullen *et. al.*.

There are some preventative measures in place to prevent land degradation processes from arable agriculture. In the Republic of Ireland, farmers protect vulnerable tillage soils by complying with 'good agricultural and environmental conditions' (GAEC) guidelines as a condition for receipt of the area-based single farm payment under the EU cross-compliance regime. Land that has been in continuous tillage for six years or more must be tested for organic matter content as a requirement for the single payment scheme. Soils having less than 3.4 per cent SOM may require remedial action depending on soil type. As the process of building up SOM is very slow, the remedial action to be taken is set out over a 10-years period. The remedial action will continue until such time as the organic matter levels are shown to have recovered to greater than 3.4 per cent or a level deemed acceptable for that soil type. Hackett *et. al.* provide information on how various management practices affect SOC dynamics in arable soils. Land application of fertilizer and manures is now subject to 'closed periods' that coincide with the most frequent average occurrence of transport vectors. Farmers are also prohibited from applying fertilizers in close proximity to a water course. 'Buffer strips' of 2m and 5m for mineral fertilizer and organic fertilizers, respectively, must be observed. The effectiveness of these aspects of the regulations is currently being monitored in the Agricultural Catchments Programme (ACP).

Soil data currently available in Ireland exists in variable forms and is not fully mapped at the target European scale (1 : 250,000). Spatial soil mapping, combined with conservative ground-truthing, is currently underway in the form of the Irish Soil Information System (ISIS).

This aims to complete the soil map of Ireland at a 1 : 250,000 scale. In a review of strategies to improve soil conservation in Europe, Fullen *et. al.* identified several best management practices including : initiation of national soil conservation services; and full mapping, monitoring and costing of erosion risk by national soil survey organisations. If the SFD is eventually ratified, Ireland will be required to identify areas where erosion has occurred in the past or is likely to occur in the future. At that time, the soil information provided by the ISIS will be essential in identifying these areas.

FUTURE INVESTIGATE DIRECTION IN THE QUANTIFICATION OF NUTRIENT AND SEDIMENT LOSS FROM IRISH TILLAGE SOILS

Sediment Provenance

Traditional techniques, aimed at identifying the source and the pathway of the sediment, have included methods such as risk assessments, field observation and mapping, landowner questionnaires, remote sensing, use of erosion pins, and terrestrial photogrammetry.

Given the time and cost involved in establishing and operating plot experiments, and that data available from them is limited, attention has been directed to the use of environmental radionuclides for documenting erosion rates. By comparing the fallout radionuclide Caesium-137 (^{137}Cs) inventory at a particular sampling point with the reference inventory (the total ^{137}Cs activity per unit surface area for a level, stable undisturbed site), the rates of soil erosion and deposition at that point can be estimated. Measurements of ^{137}Cs and the fallout radionuclide unsupported ^{210}Pb afford a means of obtaining retrospective, medium-term (*i.e.* ca. 45 years for ^{137}Cs and up to 100 years for unsupported ^{210}Pb) estimates of both the magnitude and spatial distribution of soil redistribution rates generated by sheet and rill erosion, by means of a single site visit.

Due to its long retention time on soil particles once absorbed, ^{137}Cs ($t_{1/2}$ = 30.1 years) has the disadvantage of not being suitable for the investigation of erosion resulting from individual events occurring over short periods, and is unable to distinguish between tillage and water erosion. It can, however, be used to estimate changes in soil erosion rates associated with changes in soil management practices on cultivated land. In contrast to ^{137}Cs, ^{7}Be is short-lived with a half life of only 53 days and, as such, is ideal for estimating short-term rates and patterns of soil redistribution relating to individual events (tillage or water erosion) or short periods.

Because the radionuclides ^{137}Cs, ^{7}Be, and ^{210}Pb have different distributions in the soil profile, their measurement in eroded sediment, referred to as 'sediment fingerprinting', will determine what depth in the profile the soil was eroded from and, hence, the depth and areal extent of sheet and rill erosion can be quantified as was done in a study by Whiting *et. al.* When estimating sediment erosion rates, sediment fingerprinting has the added advantage over plot studies of identifying both the source and fate of eroded sediment, which has significant implications for the development of best management practices to address soil erosion and sediment delivery to waterways. Sediment fingerprinting correlating landuse with river sediment appears to offer a valuable alternative to direct monitoring for elucidating the provenance of SS and the relative importance of spatial zones or sub-catchments comprising larger (>500 km^2) drainage basins.

In a study of one of Northern Ireland's prime salmon rivers (the River Bush) aimed at quantifying fine sediment loads and tracing in-stream fine sediment sources using sediment fingerprinting, Evans *et. al.* were able to rank the four main agents generating those sources, which were (in order of most importance) : drainage maintenance work, bank erosion (caused by increasing flow and livestock poaching), ploughed arable land, and forestry clearfell. Ploughed arable land was found to be responsible for 36.6 per cent of the suspended load and 7.5 per cent of the bed load measured in the River Bush over a 1-year period. Evans *et. al.* commented that the most likely mechanisms for transfer of topsoil to the river channel were after ploughing prior to planting and harvesting of the crop.

The best management practices recommended for the Bush catchment to reduce sediment delivery from arable land by reducing bare ground were :

- Critical area planting on land prone to long-term soil erosion;
- Planting at appropriate times as assessed on the basis of storm forecasting; and
- Vehicle movement limited across fields prone to soil erosion.

Unfortunately, as Evans *et. al.* recognised, the 1-year period of monitoring in this project was too short to provide a reliable picture of sediment dynamics in the Bush catchment.

P Loss and STP in Tillage Soils

The relationship between STP in tillage soils and DRP concentration in run-off water needs to be adequately understood and quantified for local soils. To date, in Ireland, no study has investigated the link between STP and P loss to water from tillage soils. Guidelines presently used in Ireland are based on international findings and agronomic nutrient advice. Determination of upper critical limits for P in soil should consider both the STP necessary for economic crop-production and the STP necessary to avoid excessive P loss due to erosion, surface run-off and leaching. This is essential for the development of P management guidelines for water quality that will satisfy the requirements of the WFD. Relationships developed between run-off P and STP have been used in Europe and the USA to establish threshold STP levels above which the potential threat of eutrophication in surface waters is unacceptable.

In a study to evaluate M3-P as an agri-environmental soil P test for the Mid-Atlantic United States of America, Sims *et. al.* concluded that agronomic soil tests, such as M3-P, can be used to guide environmentally-based P recommendations, and that higher risks are clearly associated with M3-P values that are in excess of concentrations needed for economically optimum crop yields. As a result of the WFD, there is increasing pressure in Europe and Ireland to develop P-based management practices that will reduce the risk of diffuse losses from agricultural land to surface waters. Modelling of P for grassland undertaken by Schulte showed that it was possible to change the range of the target P index from 6 - 10 to 5.1 - 8 mg L^{-1} P_m while still facilitating optimum productivity and herbage quality and minimising the risk of diffuse P losses to water. Index 3 (5.1 - 8 mg L^{-1} for grassland) in the new P-index system represents a target index that is both agronomically and environmentally sustainable for all soils in Ireland. The target index for tillage crops (6 – 10 mg L^{-1}) has not changed and it is uncertain if similar work on tillage soils is necessary as the risk of diffuse P loss from them has not been quantified in Ireland.

The adoption of management measures in river basins requires the ability of river basin managers to quantify the importance of different P pathways, identify and map P risk areas with a certain spatial resolution, and estimate the effect of various management measures for changes in P losses. Limited resources and time will likely hinder the carrying out of a full P loss assessment (incorporating site characteristics and nutrient management practices) on all agricultural fields in a catchment. Therefore, in the interim, there is a need to identify a STP level, sometimes referred to as an environmental threshold, above which the improvement of P management practices should be a high priority.

Using the relationship between STP in five Irish tillage soils and the DRP released in the surface run-off, Regan *et. al.* developed a run-off dissolved phosphorus risk indicator (RDPRI) to identify the STP level above which there may be a potential threat to surface water quality. The results of this study complemented the agronomic guidelines of the Nitrates Directive, as they indicate that tillage soils may produce surface run-off P concentrations in excess of 30 $\mu g\ L^{-1}$ (the median phosphate concentration above which significant deterioration is seen in river ecosystems) if their P_m and M3-P concentrations exceed 9.5 mg L^{-1} and 67.2 mg kg^{-1}, respectively.

Catchment-Scale Research

Research that will quantify the P and sediment losses associated with arable land compared to agricultural grassland in Ireland is underway in the form of the ACP. This will provide a scientific evaluation of the effectiveness of the Nitrates Directive National Action Programme (NAP) measures over time for the major farming and environmental stakeholders in Ireland. The programme is designed to assess effectiveness of measures well before improvements are expected to translate into improved water quality of the final aquatic receptors, which in some cases may take up to 20 years. In the first stage, four catchments (2 arable and 2 grassland) were selected for studying from 1500 possible candidates using spatial multi-criteria decision analysis.

Combined, the four catchments represent the range of intensive grassland and arable agriculture interests in Ireland across a soil and physiographic gradient that defines potential risk of P and/or N transfers. A fifth catchment in a karst limestone region in the west of Ireland is also being studied. The arable catchments, having between 30 and 50 per cent arable land use in each, are located in County Louth/Cavan on intermediately drained soils and in County Wexford on well-drained soils, enabling measurement of storm-induced diffuse transfers of P and losses of N to groundwater through leaching. The ACP will focus on source, pathways and delivery of nutrients to waterways over time. At the outlet of each catchment, the following parameters will be monitored : TP, total dissolved P (TDP), total reactive P (TRP), DRP, total N, NO_3-N, turbidity, electrical conductivity, temperature, and flow rate. Particular attention will be paid to P hotspots and linking these to P loads in streams. This will facilitate the identification of areas that are vulnerable to P loss and which will require measures to reduce losses. On-site bankside nutrient analysers will enable immediate analysis of nutrients susceptible to transformation if left in sample bottles for long periods of time. The use of turbidity as a surrogate technique for the measurement of SS is also being investigated in the catchments.

CONSERVATION TILLAGE

The growth of agricultural production in the United States over the past 100 years has followed the westward expansion of population across the North American continent. In the 1930's, the Dust Bowl, which resulted from a severe long-term drought and excessive tillage, prompted the formation of the U.S. Soil Conservation Service (now the Natural Resources Conservation Service NRCS) and a concerted effort to improve soil conservation practices to preserve soil resources for future generations. There are currently 295 million acres (≈120 Mha) of cropland in the U.S. subject to some form of tillage. Significant advances in machinery and cropping systems as well socio-economic changes in U.S. agriculture and government programmes have influenced the rate of conservation tillage (CT) adaptation in the US. This paper will briefly outline the state of CT in the US with a discussion of the principal determinants governing adaptation of CT across the nation.

DEFINITIONS

- *Conservation Tillage (CT) :* Any tillage and planting system that covers 30 per cent or more of the soil surface with crop residue, after planting, to reduce soil erosion by water. Where wind erosion is the primary concern, any system that maintains at least 1,000 lb/acre (≈ 1.1 Mg/ha) of flat, small grain residue equivalent on the surface throughout the critical wind erosion period. The following define three broad classes of conservation tillage.
- *No-till or Strip-till (NT) :* A tillage/planting system where the soil is left undisturbed from harvest to planting except for nutrient injection. Planting is accomplished in a narrow seed-bed or slot created by coulters, row cleaners, row chisels or roto tillers. Weed control is accomplished primarily with herbicides. Less than 25 per cent row width disturbance is considered no-till.
- *Ridge-till (RT) :* A tillage/planting system where the soil is left undisturbed from harvest to planting except for nutrient injection. Planting is completed in a seed-bed prepared on ridges with sweeps, disk openers, coulters, or row cleaners. Residue is left on the surface between ridges. Weed control is accomplished with herbicides and when ridges are rebuilt during cultivation.
- *Mulch-till (MT) :* The soil surface is disturbed prior to planting. Tillage tools such as chisels, field cultivators, disks, sweeps or blades are used. Weed control is generally accomplished with herbicides and/or cultivation.
- *Reduced Till (RDT) :* Any tillage system that leaves 15–30 per cent residue cover after planting, or less than 500 lb/acre (≈ 0.55 Mg/ha) of small grain residue equivalent throughout the critical wind erosion period.
- *Conventional Till (CVT) :* Any tillage system that leaves less than 15 per cent residue cover after planting or less than 500 lbs/acre (<55 Mg/ha) small grain residue during the critical wind erosion period. Generally involves plowing or intensive tillage.

DETERMINANTS GOVERNING ADAPTATION OF CT SYSTEMS IN THE U.S.

Socio-economic Forces

Food Security Act of 1985 : Conservation tillage was only practiced on 2.3 per cent of harvested cropland nationwide in 1965 (5). Although this percentage increased to 16 per cent in 1979 the rate of increase was stagnant. The passage of the Food Security Act by Congress in 1985 tied soil conservation practices to farmer eligibility for government sponsored crop deficiency payments, crop loans, storage payments, Federal Crop Insurance and disaster payments (4). The overall purpose of the act is to remove incentives for persons to produce crops on highly erodible land (HEL). Highly erodible land is defined as soils that would erode at eight times that soil's tolerance (T) value based on the universal soil loss equation (USLE) or the wind erosion equation (WEQ). This massive programme affects more than 125 million acres (≈ 50 Mha) nationwide. The current nationwide percentage of cropland under CT stands at 37.2 per cent, RDT at 25.8 per cent with CVT reduced to 36.2 per cent of cultivated lands.(5).

- *Farm Size* : The number of farms in the U.S. has declined from a high of 6.8 million in 1935 to 1.9 million today (11). At the same time, farm population has declined from 33 million people to under 4.8 million or 1.9 per cent of the national population. Average farm size in the U.S. today is 508 acres (206 ha). The reduction in farm labour and increase in farm size requires more timely tillage operation and less time with machinery management. Where plowing required an average of 1.22 hrs/acre (3 hrs/ha), NT requires only 0.5 hrs/acre (1.2 hrs/ha) because of the reduction in primary and secondary tillage operations.
- *Environmental Awareness* : Public awareness of environmental issues has increased throughout the U.S. as information transfer has become more sophisticated. Water quality is a principal public concern and there is increasing public financial support for conservation programmes to reduce non-point pollution of surface and ground water. Nearly every major agricultural state has some tax support for cost sharing of conservation structures on private land. Protection of watersheds with CT and forested buffer strips along the nation's watersheds is a part of national policy. Significant progress has been made towards protecting the Mississippi River watersheds in the Central U.S. however sedimentation of the nations waterways is still a major offsite effect of soil erosion which costs the nation more than $7 billion dollars annually.

Climate and Soil Factors

Agriculture in the U.S. is practiced under a wide range of climatic regimes with agricultural eco-regions that span a mix of rainfall patterns, solar energy load, day lengths, altitudes, crop requirements, soil properties, irrigation water availability/quality and pest problems. The national pattern of CT adaptation tends to roughly follow latitudinal trends with growing season length, energy flux and crop heat requirements dominating regional acceptance of high surface residue CT systems. Several of the more dominant state variables in the Central U.S. that inherently influence tillage adaptation are :

- *Soil temperature* : Seed germination and growth (especially maise) is reduced under cool soil temperatures. Radical elongation and phenology is delayed under cool soil conditions (1, 2) which reduces yield potential and increases the risk of yield loss at latitudes where heat units are marginally adequate to achieve maise maturity. The inherent characteristics of conservation tillage likely to influence soil temperature are surface residue reflectance (albedo), thermal conductivity and soil drainage class.

The relative popularity of ridge till and strip till systems in northern latitudes is attributable to increased aspect (in RT) and reduced albedo.

- *Soil Drainage :* Since the principal component of soil heat capacity is water content, CT systems that improve energy absorption in poorly drained soils through reduced albedo or increased aspect will improve conditions for seed germination and timing of spring planting operations. Conversely, soils that have low water storage capacity have improved water retention and are aided by high surface residue CT systems. Poorly drained soils susceptible to compaction are especially vulnerable to the high axel loads associated with heavier CT equipment. Even under NT, 30 to 50 per cent of the soil surface may be subjected to wheel traffic.
- *Water Conservation in Fallow Tillage Systems :* Dryland fallow systems are practiced in the High Plains of the western USA where two wheat crops are grown in three cropping season. During the lengthy fallow period (≈14 mo.) soil is managed for weed control, water harvest, snow capture and wind erosion protection. Tillage systems that result in erect stubble and high surface residue with a greater reliance on herbicides for weed control have dramatically increased fallow efficiencies. However, economic return to land, labour, capital, and management in fallow systems are reduced under no-till as the cost of conversion to no-till from stubble mulch is greater than the yield advantage from increase fallow efficiency. Costs for NT are $8 - $11.00 more per acre than stubble mulch (6,10). Intensification of cropping towards shorter fallow periods, however, can improve profitability under NT.
- *Water and Wind Erosion Hazard :* Obviously conditions that reduce run-off and improve water infiltration will reduce water erosion potential. Conservation tillage practices tend to increase the tortuosity of surface run-off path and increase infiltration time. Conservation tillage systems dominate regions of the country where farming is practiced on sloping lands, soil texture is conducive to crusting and erosive rains are frequent during soil preparation for planting. Wind erosion is the dominant hazard in the wheat-fallow regions of the Central Plains and CT systems that preserve erect residue will reduce wind velocity at the soil surface. At times, tillage operations are performed to increase soil roughness and aggregate size to reduce saltation under extreme conditions.

Crop Management Factors

Weed Control : Conservation tillage systems most often result in increased reliance on chemical weed control. Weed control can be a significant problem in CT systems where pH stratification affects herbicide activity or where residue cover may intercept herbicides rendering them inactive. At times, emergency tillage is required to offset herbicide failures resulting in less than 30 per cent residue cover. In specific situations, the most effective herbicide compounds will require soil mixing limiting the choice of CT systems. Conversely, CT systems can also afford reduction in herbicide use when banded in strip- or ridge-tillage operations. Cooler soils in residue-laden zones may suppress the emergence and competitive ability of weeds relative to the crop. Weed control in continuous cropping systems are usually the most problematic for CT systems because weeds with similar biological characteristics and growth periods to the crop tend to predominate over time. In these situations, rotation of crop species may be the most effective weed control measure under CT. The introduction of herbicide resistant crops has revolutionised CT weed control strategies in the maise belt but the safety and ecological impact of these crops is being questioned.

Crop Rotation : Crop rotations allow selection of different herbicides, planting dates and tillage systems resulting in greater opportunities to control weeds, disease and insect pests. Certain CT systems are not amenable to crop rotations. For example, RT systems that require ridge construction during a row crop season are unsuited to rotation with solid seeded or sod crops. This fact tends to limit the acreage devoted to RT. Often CT systems alternate with crop rotations in response to residue management needs or fertilization strategies. Crop rotations of all sorts are becoming the dominant cropping strategy as they provide market diversity as well as management flexibility.

Soil Fertility : Stratification of soil chemical properties and nutrient distribution is a commonly observed phenomenon of CT systems. Non-mobile nutrients such as P and K can accumulate in the upper portions of the soil profile from surface placement of fertilizer materials and/or deposition from decomposed residues. The extent of soil acidification produced by nitrification of ammonium N will depend on soil buffering capacity and inherent N mineralisation potential. Re-distribution and mixing of nutrients varies with the degree of soil mixing so the choice of CT system and duration of CT practice will depend on native soil fertility and the resistance of soil to chemical change.

TILLAGE SYSTEMS FOR SOIL AND WATER CONSERVATION

The care of soil is essential for the survival of the human race. It provides most of the food required, fibres for clothing and wood for building materials. But in many parts of the world the soil has been so damaged by mismanagement that it will never be able to produce, at least not in the near future. This maltreatment has led to soil erosion on a vast scale, that is, physical loss of upper layer of soil which is most vital for plant nutrition. Severe soil erosion has taken place even in advanced countries. It has become a global problem and its remedy lies in the application of soil and water conservation practices through a soil conservation programme. However, the farmers themselves have to carry out the necessary practices on their land as a part of their farming system.

In the early stages of the development of crop culture, vegetation that surrounded and competed with plants (that provide food) was flattened and removed by hands. Food production was generally limited under these conditions. As man learned to grow certain food plants in the more desirable locations, it became necessary to remove existing vegetation which heralded the beginning of soil tillage.

For tilling the soil in older days, crop producers used crude implements made of wood or stone. With the help of the implements, they used to loosen the soil but gradually they realised that weeds are primarily responsible for restricted growth of plants. Thus, improved implements and techniques for weed control have been developed. Weed control is one of the basic reasons usually given for tilling a soil.

In the beginning the agricultural practices were simpler and the tillage operations were also simple. But in the modem agriculture, tillage is defined and described as follows. The operation of implements through the soil to prepare seed-beds and root beds, control weeds, aerate soil and cause faster break- down of organic matter and mineral to release plant nutrients.

Types of Tillage

Different types of tillage systems are given here :

- *Clean tillage :* Cultivation of a field so as to bury all plant residues and to prevent growth of all vegetation except that of the desired crop.
- *Conservation tillage :* Any tillage sequence that reduces soil or water loss relative to conventional tillage. It is often a form of non-inversion tillage that retains protective amounts of crop residues on the surface.
- *Conventional tillage :* The combined primary and secondary tillage operations normally performed in preparing a seed-bed for a given crop in a given geographical area.
- *Minimum tillage :* The minimum soil manipulation necessary for crop-production or meeting tillage requirements under existing soil and climatic conditions,
- *No tillage :* A method of planting crops that involves no seed-bed preparation other than opening the soil for the purpose of placing seed at tile intended depth; usually involves opening a small slit or punching a hole in soil; usually involves no cultivation during the growing season; usually involves chemical use for weed control, Also called as slot planting, zero tillage, direct drilling,
- *Reduced tillage :* A system in which the primary tillage operation is performed in conjunction with special planting procedures to reduce or eliminate secondary tillage operations; less than conventional tillage. Similar to minimum tillage, sometimes called limited tillage.
- *Stubble mulch tillage :* A system of tillage that retains the stubble of crops or crop residues in place on the land, thus providing a protective surface cover before and during seed-bed preparation and at least partially during the growing season of the succeeding crop,

Objectives of Tillage

Although soils are tilled for controlling weeds and also for other reasons, the very goal of tillage is to stabilise and increase crop-production. Tillage is the physical manipulation of soil with tools and implements to result in good tilth for better germination of seeds and subsequent growth of crops. In other words, it is the tilling of land which is done for bringing about conditions that are favourable for the cultivation of crops. Both implies to the physical condition of the soil in relation to plant growth. Tilth is brought by tillage.

Tillage has got several objectives :

- The main objective of tillage is the preparation of land to make it suitable for crop cultivation. It makes the soil loose and friable with sufficient aeration and moisture for the germination of seed.
- Tillage helps 10 control weeds and incorporate the weeds, crop residues, green manures and other organic manures and fertilizers, soil amendments and other agro chemicals applied for the control of weeds, insect pests etc.
- It improves soil structure, soil aeration, water holding capacity of soil, soil permeability, organic matter decomposition and creates suitable conditions for root penetration removing hard pans, if any, in the soil.
- The most important objective is to prepare good seed-bed.
- Proper tillage results in soil and moisture conservation through high filtration, reduced run-off and increased depth of soil for moisture storage.

To emphasise upon the soil and water conservation, the need of the hour, one has to adapt tillage systems that are most suitable for the prevailing soil, crop and climatic conditions; Under some conditions a simple change in tillage method may be adequate.

Selection of Tillage Systems

The tillage system selected for a particular situation depends on climatic zone, crop to be grown, soil factors, economic level of the producer, preference of the producer etc.

Climatic Factors

The climatic factors like rainfall, temperature, radiation and wind are the most important ones while considering the tillage system to be adopted.

Rainfall

In areas where rainfall is low, tillage systems and related practices for the conservation of water are highly desirable because as much moisture as possible must be stored in soil for the use of plants for their growth. In order to achieve this, tillage systems which can improve the water infiltration rate, can trap moisture from snow and dew and can reduce the evaporation from the soil surface should be adopted.

Infiltration can be enhanced by reducing run-off rates, making the soil surface condition suitable for rapid infiltration and removing soil profile layers which restrict the water movement down into the soil. Reduced evaporation can be achieved by improving the micro-climate at the soil air inter-phase thus increasing the storage of water in the soil. Mulching not only helps in improving the micro-climate but also intercepts the radiation, avoid high soil temperature and make the soil rough which in turn reduce the wind speeds.

In scanty rainfall areas, water and wind erosion are the major problems which can be controlled by the practices mentioned above.

In regions where rainfall is high, use of grassed waterways is desirable as they conduct excess water away from the fields.

Temperature

The prevailing temperatures strongly influence crop adaptation in a region. Tillage systems have little or no effect on air temperatures, but can be used to manipulate the soil. This should be particularly useful in areas where frost-free periods are relatively short and need few more extra days of growth period. Mulching will help in maintaining the soil temperatures. A surface mulch decreases day time soil temperature but a combination of mulching and ridging in the field gives the effect of a tillage and can maintain soil temperature and soil water content and also improves the use of conserved water, thus protects soil from erosion.

Radiation

Tillage affects the radiation balance of a soil mainly through its effect on soil colour and water content, slope relative to the sun and surface residues. Light coloured surfaces reflect radiation whereas dark ones absorb it. Thus, a freshly tilled, rough moist soil will absorb more radiation than a smooth, dry soil. On sloping soils, absorption is greatest when the sun is perpendicular to the slope.

The radiation balance due to surface residues is affected by the colour, decomposed stage and geometry of residues. Bright residues reflect a large percentage of the incoming radiation. As residues decompose and darken, reflectance decreases. The radiation balance has a direct effect on photosynthesis. It also indirectly affects water conservation through its effect on soil temperature.

Wind

Wind speed has a major influence on soil. It is advised to take up cultivation or tillage systems suitable for a particular climatic zone mainly because of the potential of different tillage systems for controlling wind erosion.

On soils subject to wind erosion, tillage systems that maintain surface residues or provide a rough, cloddy or ridged surface during the windy season should be used. Where winds from one direction predominate, the direction of tillage operations that ridge the soil surface should be perpendicular to the prevailing winds. On soils where the wind erosion potential is much greater than water erosion potential, soil ridging tillage should be perpendicular to the prevailing winds than on the contour. While contour tillage would control water erosion, wind blowing parallel to the ridges could cause greater erosion than that caused by water during infrequent rainstorms. This would be especially true on deep sandy soils having relatively high infiltration rates.

Crops to be Grown

Different tillage systems are adopted for different crops. Differential crops responses are mainly due to the tillage systems adopted. They affect soil water content, weed control, aeration, root zone depth, fertility etc. Some crops like fibrous rooted grain crops are not affected by the soil physical conditions resulting from tillage methods but their growth is related to the amount of water available in the soil. In contrast, root crops such as sugar beet, responses to tillage induced changes such as water infiltration rate, aeration, bulk density changes etc.

Tillage systems that provide a specific soil condition may also be required for some other crops *i.e.,* for tuber and root crops a deep loose root-zone is required and small seeded crops requires uniform finely granulated seed-bed. Apart from these, short statured crops like vegetables and cotton, trash free surface is essential as they may interfere with harvesting operations or lower crop quality.

Soil Factors

Soil factors like slope, texture, depth, density, salt or alkali content and drainage are important in selecting the tillage systems.

Slope

On nearly level or gently sloping soils, many tillage systems provide the desired conditions and effectively conserve the soil and water resources. As slope increases the choices of tillage systems become more limited. A conservation tillage system that maintains surface residues is best suited for water erosion control on sloping soils when other supporting practices are not used. When supporting practices such as contouring, terracing and strip cropping are used, tillage is possible on soils of somewhat greater latitudes. Land leveling permits the use of almost any type of tillage systems. However, on steeply sloping soils, the levelled areas become extremely narrow and therefore the use of some tillage methods or systems may be impractical. Tillage may even be restricted to use of hand implements on such areas.

Texture

Soil texture has a major influence on a soil's susceptibility to erosion and therefore, on the selection of tillage methods or systems for controlling erosion. Conservation tillage systems, as a rule, are highly effective for controlling wind and water erosion on all types of surface textures.

Sandy soils, in which very less crop residues are available, are susceptible to wind erosion. Normally chisel plough is used on these soils than mould board or disc plough. To be effective the tillage should be performed perpendicular to the prevailing winds and at a soil-water Content conducive for forming a rough surface. Clods should not be pulverised. Mould board ploughing is sometimes used on sandy soils to bring finer textured material to the surface to aid in forming clods and, thereby reduce susceptibility for erosion.

The choice of tillage method or system is greater on finer textured soils for controlling wind erosion. The major pre-requisite of any tillage system for controlling wind erosion where surface residues are not available is to produce a rough, cloddy or ridged surface. About 75 per cent of the surface soil material should be greater than 0.84 mm in diameter to reduce the soil losses due to wind erosion.

Effective control of water erosion usually depends on the use of supporting practices such as contouring, terracing, strip cropping and crop rotations in conjunction with the tillage system. Contouring and terracing reduce slope gradient and length and reduces the loss by run-off. Strip cropping, depending on the crop in the strip reduces run-off velocity and traps the sediments. The land area, where crop rotation has been practiced constantly, would become somewhat resistant to erosion because, there would be less soil dispersion, less surface sealing and greater water infiltration and therefore, less run-off and erosion.

Sometimes the tillage system itself may decrease the soil residues. In general, best control is obtained with tillage that maintains unsealed soil surface and permits high water infiltration rates. For this purpose presence of water-stable aggregates are desirable which normally result from maintaining, as much organic material as possible at or on the soil surface. When precipitation rates exceed infiltration rates, temporary surface storage of water reduces run-off and aids in erosion control.

Ridge forming tillage on the contour is a long proven run-off control and water conservation practice. Run-off can be eliminated on the contour of gently sloping land by adopting lister ploughing (lister plough is a plough with double mould board) in conjunction with closed-end level terraces or by basin listing. For subsequent temporary storage on the surface, water collected behind the ridges or dams must infiltrate before the next precipitation (rainfall) event.

In sandy soils, by increasing the clay content water holding capacity can be increased. If too much clay is placed in the surface layers, it may decrease the infiltration rate. However, at least 8 per cent clay is needed in the surface layer of sandy soils for subsequent tillage to result in clods that resist wind erosion.

Depth

Selection of tillage systems based on soil depth is mainly concerned with the depth to an untillable layer (like bed rock) or depth to a layer that would contribute undesirable substances like sand, gravel, rocks, high calcium materials, saline or alkali materials or strongly acid materials etc., to the till- age zone.

On shallow soils, non-inverting tillage should be used to minimise damage to equipment or the danger of contaminating surface soil with the undesirable substances. Rolling type equipment such as disc harrows, ploughs, disc openers planters are generally well adapted for use on shallow soils underlain by rocks. However, disc tillage is not desirable because it accelerates erosion. Under shallow soil conditions, a conservation tillage system involving minimum soil distrurbance, such as use of disc opener planter and applying herbicides for weed control would be most desirable.

Any tillage system that provides desired soil condition and conserves soil and water resources can be used on deep soils.

Dense Horizons in the Soil Profile

If there are any compact layers in the profile they affect the water infiltration rate. These compact, restricting layers like hard pan, fragipan, plough sole etc. may be formed from tractor and implement operations or animal trampling or can be formed naturally also.

Hard pans having rock layers, generally are not disrupted. However, fragipans, plough soles etc., can be broken by chiselling or by doing deeper than normal ploughing. Response to those operations would be greater when performed with the soil relatively dry. Dense or compact zones when disrupted increase water infiltration, conservation, greater root penetration and proliferation to extract water from the soil.

Alkali or Salt Content

Alkali soils can be reclaimed by adding gypsum to the soil which usually gives the greatest benefit with respect to water infiltration, because harmful salts would be leached down.

Apart from all these factors economic level of farmer to adopt tillage systems and his reference should also be considered.

Tillage Systems

Since man began to cultivate the soil many tillage systems have been come into practice of which important ones are described here.

Shifting Cultivation

This is a widely practiced system in different parts of the country. Major advantages of this system are the low capital input requirement, dependence on natural or regenerated soil fertility and the opportunity to grow a wide variety of crops. As it involves clearing of the forest land by fire, it destroys weed seeds and to some extent weed control is being achieved. But non- application of fertilizers and herbicides results in declining soil fertility and greater weed problems. Also burning of forest litter, crop residues, grasses etc., may control the weeds but make the soil susceptible to erosion. As the fertility declines, there is a need to change the plot in every 2 to 5 years.

Labour Intensive Continuous Cultivation

Continuous cultivation refers to annual cropping where one or two years of occassional fallowing may be there unlike shifting cultivation in which land is fallowed for a longer period than the cropping period.

There are two systems that comes under this cultivation :

1. *Subsistence continuous cultivation :* Labour intensive continuous cultivation at the subsistence level is similar to shifting cultivation in most regards except that a plot is cropped for a longer period than it is fallowed. Farmers mainly produce sufficient quantity for their family. Because a plot is cropped continuously major clearing is not necessary in each season. However, greater. effort is usually required to control weeds because weed problems are generally increased with cropping period.

The basic tools used in this system are the hoe and machete (a large heavy knife) or cutlass (a short, heavy slightly curved knife).

Under poorly managed conditions with no provisions for soil fertility maintenance, and soil and water conservation, crop yields decrease and land degradation intensifies. With improved management, such as soil fertility maintenance through recycling of various organic wastes, growing soil improving crops in rotation, use of other resource conserving practices etc., good crop yields can be obtained without serious land degradation.

In order to maintain the productivity, fertilizer application, use of mulches, adopting soil and water conservation practices, can be put into practice.

2. *Extra production continuous cultivation :* This system is also based on hand labour for the necessary cultural operations. The basic tools are spades, hoes and also some other implements are used.

Under this system farmers normally produce more than they require for their families. They also grow one or two crops at commercial level. Animal production may also be involved in association with crop-production.

Advantages of extra-production continuous cultivation are; conservation of soil and water resources, improves soil fertility, recycling of organic wates, and improve soil physical condition. Recycling of organic wastes is labour intensive. Mixing residues with soil by use of hand implements is also difficult. Therefore, the development of no tillage systems for arable crop-production is practiced for efficient recycling of organic wastes. Under no tillage system of cultivation, soil loss is less as it involves minimum disturbance to the soil. Maintenance of organic matter is also undertaken as it increases the water holding capacity, infiltration into the soil.

Major disadvantages are high labour requirements, limited arable land for expansion, limited organic wastes to manage, costly or limited fertilizers etc.

In labour intensive continuous cultivation systems, hoe cutlass or machette and spade are the main implements used.

Hoes are the major weed control implements and hoeing helps in smoothening the surface, breaking and pulverising the clods and spades can also be used in loosening and mixing the soil.

Animal-drought and Small Tractor Cultivation

The era of animal draught technology for agriculture began about 3000 B.C. when man learned to harness the animals he had domesticated earlier. The wheel was discovered which eventually lead to the development of carts pulled by humans and animals and ultimately tractors were developed for tillage operations. Essentially any soil condition achieved by use of animal or tractor power can be achieved by hand labour. But the major difference lies in the labour required, timeliness of the operations and intensity with which soil is tilled.

Once the soil is tilled for controlling the weeds the moisture may get evaporated from the soil. As for weed control, timeliness of seed-bed preparation also can have a major effect on water conservation. Seed-beds should be ready whenever conditions are optimum for establishment of the given crop. This can be achieved by performing major tillage operations for seed-bed preparation well before the optimum planting date. This may require that the tillage be performed under relatively dry soil conditions, especially in regions having distinct low and high rainfall, regions. At some locations, light showers during the dry season are beneficial for seed-bed preparation before onset of the rainy season. Final seed-bed preparation can then be achieved without major delay at the start of the rainy season, thus resulting in more timely crop establishment and potential for using growing season rainfall more efficiently for crop-production. Such early seed-bed preparation and crop establishment is conducive to water conservation in several ways. So, early tillage helps in water conservation.

In this system, the need for capital increases. When the cultivation system is subsistence or labour intensive cultivation system, the need for capital is low, because most of the labour is supplied by the farmer or family members and the implements used are simple. The need for capital increases with the introduction of animals or small tractors into the cultivation systems. Capital is required by the farmer to acquire the animals or tractor, satisfactory implements, to provide food for the animals and fuel for the tractor and to cover other production expenses.

Modem High Technology Cultivation

Modem high technology refer to those systems in which most crop-production operations are performed with fuel powered machines like tractors (include harvesters, balers etc.) and other implements and machines, with the aid of chemicals (herbicides, insecticides etc.) Labour is limited mainly because many operations are carried away by the machines and manual labour may be necessary for some crops to reduce plant population and for weeding and harvesting. In most of the cases, crops are planted at a proper seed rate to give desired plant population and weeds are controlled by applying herbicides. For harvesting of crops also mechanical means are used but hand labour is still widely used and even necessary for harvesting some crops.

A major advantage of this system is that field operations can be accomplished rapidly when conditions become favourable. For example, family operations like ploughing, crop establishment and harvesting can be completed quickly when conditions are optimum. But energy and equipment requirements are high in this system. Chemicals play an important role and are widely used to control weeds, diseases and insects. Under favourable plant and climatic conditions, one or a few correctly made applications of herbicides can effectively control weeds that would normally require one or more tillage operations and in many cases, many additional hours of hand labour. For example, in wheat-fallow-sorghum cropping system, three to four tillage operations are required for weed control.

CLEAN TILLAGE SYSTEMS

Clean tillage systems are those in which all plant residues are covered and in which the growth of all vegetation is prevented except for the desired crop. The residues are usually covered by inversion type tillage early in the interval between crops. In some systems, however, residues may be par tially incorporated with soil at the first operation, as with disking, then further incorporated at subsequent operations so that little or no residues remain on the surface when the next crop is planted. Unwanted vegetation is controlled initially by the major ploughing operation and subsequently by one or more forms of secondary tillage such as disk harrow, chisselling, sweep ploughing etc. During crop growth period, weeds may be controlled by cultivation, hoeing or application of herbicides.

As a general rule, clean tillage systems are adaptable and suitable on lands (please refer booklet No. 518 on Land Capability Classification) that have few limitations for crop-production. Clean tillage systems may be adaptable to other classes of land (Class II and, III), provided appropriate supporting conservation measures are used and the choice of plants is restricted to those that provide adequate protection against soil and water losses.

Most, if not all crops are adaptable to clean tillage systems on class land. Clean tillage is especially appropriate for crops requiring well prepared uniform seedbeds and precise planting such as many seeded vegetable crops. This, however, does not mean or imply that clean tillage is the best system for all crops even when produced on class land. For some crops, precise planting in a residue free seed-bed is not essential and surface residues retained by other types of tillage may enhance crop-production through greater water conservation, less surface crusting and improved seedling establishments and plant growth as compared with that obtained with clean tillage.

The threat of erosion on class land is slight and therefore surface residues are seldom required to control erosion on such land. However, other classes of land, surface residues retained by other forms of tillage provide some or full protection against erosion which cannot be achieved by use of clean tillage.

Clean tillage is achieved through a variety of implements such as mould board, lister and disk ploughs, which eliminate most or all crop residues at the first operation. Other implements such as disk harrows, chisels and cultivators incorporate some residues at each Operation.

In conditions where a clean seed-bed is desirable at planting, but where soils are particularly subject to erosion during a part of the interval between crops, use of residue-conserving tillage early in the non-cropped period, followed by residue incorporating tillage late in the period may produce the de- sired results. For example, chisel, sweep or disk tillage could be used to maintain a large portion of crop residues on the surface during an erosive period that occurs early in the interval between crops. Later as planting time approaches, disking or lister ploughing may provide desired seed-bed conditions.

Mould-board or disk ploughing is seldom desirable as planting time approaches because one or more secondary operations may be required to obtain a proper seed-bed. In addition, major tillage near planting time could result in high evaporative losses of water from the tillage layer, thus reducing me amount retained in soil for crop use.

Thus, where the clean tillage systems, have been adopted on those soils little or no residues are available tor incorporation, but tillage induced surface roughness and cloddiness can increase water infiltration, reduce run-off velocity and thereby reduce the potential for soil loss. Additional storage volume is created by tillage but it is a temporary one. So it is advised to adopt conservation tillage systems in order to increase the water storing capacity of the soil. There are four major conservation tillage systems and they are described in the succeeding chapter.

Advantages of Clean Tillage Systems

In this system, less troublesome performance of cultural operations (like tillage, planting spraying etc.) improved weed control, improved crop establishment, better insect and disease control, greater soil-nutrient interaction and higher crop yields are obtained.

Disadvantages

Disadvantages of clean tillage systems with respect to soil and water conservation result primarily from conditions at the soil surface. However, sub-surface conditions namely soil compaction and a general decline in organic matter content are also involved. A further disadvantage is that it is an energy intensive system.

Conservation Tillage Systems

Conservation tillage systems are systems for managing crop residues on the soil surface with reduced or no tillage. The reductions may be with respect to frequency and intensity of tillage. Various types of conservation tillage systems are discussed here.

Stubble Mulch Tillage

Stubble mulch tillage refers to tilling of soil in such a way that plant residues or other materials are maintained to cover the soil surface. It is also referred to as mulch farming, trash farming, mulch tillage and plough-less farming. In a broad sense, any form of tillage that results in plant residues being maintained on the soil surface could be classified as stubble mulch tillage. Disk type or other implements incorporate some residues into the soil when unusually large amounts are retained on the surface and leave some on the surface to reduce erosion. Primary tillage is done generally 12-15 crns deep. Subsequent tillage is done at successively shallower depths. This tillage is adaptable to all types of soils. However, coarse textured soils require more surface residues to control wind erosion than fine textured ones. Therefore, wind erosion is more easily controlled with stubble mulch tillage on fine textured soils than on coarse textured ones. Soil erosion by water is easily controlled in coarse textured soils because water infiltration is rapid in them. So, less surface residues would be required on coarse textured soils to control erosion by water.

The stubble mulch is better adapted to arid and semi-arid climatic regions than humid and sub-humid regions. In dry areas it lowers the nitrification, thereby prevents over-stimulation of plant growth thus improves water fertility balance. Improved weed control is also observed. The rate of organic matter decline is low in this tillage system.

A problem sometimes encountered with stubble mulch and other conservation tillage systems is the toxic effect of substances from residues on subsequent crops; which is know as phytoxicity or allelopathy. In humid and sub-humid regions, problems arise probably due to phytoxicity only. However, for effective soil and water conservation, sufficient residues should be maintained on the soil surface.

Minimum or Reduced Tillage

Controlling weeds is a major reason for tilling a soil. Therefore, if weeds can be controlled by another means, the need for tillage is reduced and such an alternative is to use herbicides. When such alternative is adopted, the minimum tillage can be defined as minimum soil manipulation necessary for crop-production under existing soil and climatic conditions.

Major types of minimum or reduced tillage systems are given here :

- *Autumn plough :* In this system, primary tillage is With the mould board plough and secondary tillage is reduced to one shallow cultivation which is followed by leveling the soil by planking at the time of planting. A disk or rotary tiller may be used instead of the field cultivator to produce finer, firmer seed-bed. It is widely used on the dark, nearly level, medium to fine textured clay loam soils.

- *Spring plough :* In this system, strip seed-bed is prepared on soil that was initially ploughed *i.e.* only 12-24 hours before planting the crop. By planting soon after ploughing, the soil does not dry appreciably and the soil water content at planting, would be such that the clods get broken and make the seed-bed firm. This system provide greater protection against erosion because residues from the previous crop are maintained on surface until ploughing.
- *Disk and plant :* The initial disking is done usually in the autumn followed by one or more diskings in spring before planting.
- *Strip tillage :* Only a narrow band of soil is tilled in this system. A typical tillage zone is about 20 cms wide and 5-10 cms deep.

In this minimum tillage system, crop residues are maintained for a longer time or throughout the crop season. Improved soil and 'water conservation is definite advantage of this system. Through this effective conservation, productivity of land can be maintained for sustained crop-production. As it involves fewer cultural operations or less intensive soil manipulation, it reduces labour expenses, tractor and equipment usage, time and fuel energy is also reduced and therefore result in more economical crop-production. It maintains organic matter in the soil. By residue decomposition nutrients would be liberated and the loss by leaching, volatalisation and erosion is reduced.

Some disadvantages are, lowers soil temperature, pest problems, poor seed placement and lower nutrient availability.

No Tillage

No tillage is a method of planting crops that involves no seed-bed preparation other than opening the Soil for the purpose of placing seed at the desired depth. This involves opening a small slit of punching a hole in the soil to place the seed. Usually the crops are not cultivated and chemicals are normally used to control weeds. For practical purposes to facilitate crop planting, up to 25 per cent of the surface area can be disturbed or tilled in a no tillage system. One shallow disking (not deeper than 7.5 cms) is sometimes necessary to establish a cover crop or to cut crop residues in no tillage systems.

Advantages of no-tillage system include improved control of wind and water erosion, increased use of land, improved water conservation, reduced energy requirements, reduced labour requirements and equipment inventories, reduced wear and tear on tractors and equipment and greater net returns.

Disadvantages include limited supply of equipment for the application of herbicides and water for diluting the herbicides.

Dust Mulches

Dust mulching is essentially a clean tillage system that could be used in any of the major cultivation systems. It consists of loose, finely granular or powdery soil at the soil surface and is usually produced by shallow tillage or cultivation. Water conservation with a dust mulch is greater than barren lands hut when compared with other systems, it is not that much effective.

Equipment for Conservation Tillage Systems

Under the stubble mulch tillage systems machines with blades sweeps are used for primary tillage. Subsequent or secondary tillage operations are performed with a sweep machine, spring tooth cultivator, chisel plough etc. The last tillage operation can be carried out by rod weeder which does shallow ploughing and provide a firm seed-bed. Fertilizers can be applied with chisel applicators. Drills with hoe openers are widely used for seeding. Weed control can be done with herbicides.

In minimum or reduced tillage, various types of medium to large tractors can be used. Primary tillage can be carried out by mould board plough or disk plough or chisel plough, rotary tiller etc. Planting and fertilizer application is similar to that of stubble mulch system. Weed control is done by herbicidal application or cultivation or by hoeing. In no tillage systems, a herbicide applicator, fertilizer applicators, a seeding unit and a power source are required. A harvester capable of chopping or uniformly spreading crop residues is also needed. For applying herbicides, knapsack sprayer is widely used. Drills are required for the placement of seeds and fertilizers.

THE ECONOMICS OF CONSERVATION AND CONSERVATION TILLAGE

At first glance, any conservation measure such as conservation tillage can be viewed as simply another production decision for the producer : "Will it improve my profits?" However, the purposes of changes in agricultural production practices generally fall into at least one of three categories :

1. To achieve improved profitability, a private economic decision for the individual producer and landowner;
2. To satisfy a lifestyle of the producer, a private non-economic decision for the individual producer;
3. And/or to improve or preserve ecosystem characteristics, a public decision with possible benefits to society, and benefits or costs or both to the individual producer and landowner.

Depending on the situation, conservation tillage may fall into all three of these categories. This paper will briefly review the concept of conservation tillage as practiced in the United States in recent years, provide background on the cultural and political economy of soil conservation and tillage practices, summarise selected studies regarding the economics of adopting such practices, and suggest areas for further study.

Background

The value of and need for soil conservation practices has been an important part of the agricultural history in the United States. As settlers moved west and south, they often found through bitter trial and error that many soils were too fragile for intensive production practices. Following on the heels of the Dust Bowl of the early 20th Century, research by public institutions and private producers resulted in the development of a variety of conservation practices. Over the past few decades, refinements have been made in conservation tillage practices. One researcher has suggested that changes in tillage systems resulted from changes in machinery and the development of chemical herbicides.

The bottom line for the producer, however, remains the bottom line—profits. If conservation is profitable, there is a likelihood the practice will be adopted. The producer reduces or eliminates cultivation while increasing the use of chemicals such as herbicides, and yields typically suffer. The hope is the net gain will be positive. When the market provides the profit incentive, so much the better. Often, however, the market does not value such a practice and it is up to society through public programmes to provide incentives. This has occurred in various forms in the United States over time. Research, demonstration plots and extension education are examples of free public services provided as incentives. Cost-share programmes through state and federal agencies have also provided incentives.

Public programmes such as the Conservation Reserve Programme (CRP) of the federal 1986 farm act provided subsidies to actually take land out of production for 10-15 years. As such contracts have com to their end, producers have been faced with options of requalifying for new CRP contracts, returning the land to crop-production, or finding alternative uses for the land. When producers consider returning the land to crop-production, there is the obvious opportunity to move directly into conservation practices such as conservation tillage. In some cases, this may even be a necessity because of conservation compliance requirements for participation in all government programmes. It is critical that the producer evaluate the economic alternatives and consequences.

From anecdotal evidence, the decision to implement new practices such as conservation tillage is not purely an economic decision. Often, there are psychological and cultural barriers such as "the way its always been done", or peer pressure of how to farm and how a clean field should look. Another term used for conservation tillage is "trash farming", suggesting a negative, unappealing practice. Even when economic evaluation seems to support conservation tillage, consideration about the assumptions related to prices, markets, weather, new equipment purchases, soil condition and types, and skill building are critical.

Economic Factors

A sampling of studies has been reviewed and are summarised here. Most research has shown that costs such as fuel (perhaps 50 per cent) and labour (perhaps 40 per cent) can be lowered, while some costs may increase at least initially such as chemical and equipment and management costs. The primary trade-off comes with yield reduction. Studies on conservation tillage applied to cotton production in the south-eastern U.S. generally suggest little difference in yield or net revenue in the early years of establishment, but higher net revenue for conservation tillage over the long-term. In Arkansas, when yield varied, so did net revenue. Where labour saving were realised with conservation tillage in Tennessee, it was more economical than conventional tillage. A two-years study of cotton fields in Texas showed net returns for conservation tillage even when yields were lower.

Survey analysis in Oklahoma suggested that producers would lose money if adopting no-till as they returned to wheat or sorghum crop-production upon completion of a CRP contract of 10 years. Government subsidies would be necessary to make the decision pay. Another Oklahoma study over a 10 years period indicated that alternative tillage methods were less economical than moldboard plow clean tillage. In fact, the no-till system was the least economical. This study did note a limitation of the findings was that external benefits and costs were not evaluated. Continuous mono-culture winter wheat for grain with no rotation has not been found to be economical for conservation tillage, primarily because of the cost of herbicides. Some researchers do suggest that technological innovations, especially with genetically-enhanced seed, may allow significant reductions in the cost conservation tillage practices.

A Missouri sustainability study showed alternative tillage practices using ridge till to be more economical than conventional farming. The study did note additional non-valued benefits of decline in topsoil loss and supportive of a larger community economy with the ridge till. Another study in Missouri suggests that where conservation tillage does not work well, profits can be maintained by shifting out of crop-production and into livestock grasing. An Indiana study of corn and soybean production concludes that switching to no-till increases profits. This study also noted that, while leasing arrangements may complicate the conservation tillage decision, skilled producers can still benefit.

Often noted but seldom economically evaluated are the physical environmental impacts such as topsoil protection, water conservation, increased and decreased chemical use, and energy conservation. One study indicates that CO_2 loss declined as more conservation tillage takes place. Carbon sequestration may enhance environmental quality. National data suggest that residue remaining after planting has continued to increase over the past decade as more acres are submitted to conservation tillage practices.

Chapter 27

EFFECTS OF SALINITY STRESS ON GROWTH OF *CALOTROPIS PROCERA* SEEDLINGS

Hassan S. Al-Zahrani

King Abdulaziz University, Faculty of Science, Department of Biology, P. O. Box 80054, Jeddah 21589, Saudi Arabia

ABSTRACT

The effect of NaCl salinity (0, 5, 10, 20, 40, 80, 160 and 320mM) in different ages on the growth of *Calotropis procera* (Ait.) Ait. seedlings were studied. Salinity stimulates the seedling growth especially under low salinity stress. The growth of seedlings affected by high salinity stress (160 and 320 mM NaCl). In general, *Calotropis procera* seedlings can grow and tolerate low and medium NaCl salinity.

Keywords

Salinity stress, *Calotropis procera,* Seedling, NaCl salinity.

INTRODUCTION

Among the several soil environmental conditions which can limit successful growth of plants is salinity. Soil salinity is increasingly becoming a major problem of soils, it is a wide spread in many parts of the world especially in the arid and semiarid regions which has low rainfall and high evaporation (Flowers, *et al.*, 1986 and Larcher, 1995). Ecological and physiological investigations are needed to understand the behavior of the desert plants in relation to some environmental factors such as salinity. These studies are essential for their preservation since they are important ecologically, physiologically and economically.

Calotropis procera (Ait.) Ait. (milk weed) is a local wide spread shrub over abroad range of altitude in Saudi Arabia (Migahid, 1990). It is one of the most important common weeds known for the toxicity of the milk (latex) in their tissue and for its economic and medical importance.

MATERIALS AND METHODS

Seeds *of Calotropis procera* (Ait.) Ait. were collected locally from the freshly gathered ripe fruits, obtained from various shrubs growing in Jeddah area. Seeds were germinated on two layers of filter paper in flat trays, and irrigated with distilled water until they were large enough to handle (after 16 days). The seedlings were then transferred to boxes (3 liters) which had been painted black containing full culture solution (Hogland and Arnon, 1950). Seedlings were suspended above the solution by supporting them with non-absorbent cotton inside holes in small depressions in the box lid.

The plants were grown in growth cabinets with a 12-hr photoperiod, the temperature ranging between 25°C during the day and 20°C during night periods and the relative humidity of about 70%. Salinity treatments were imposed (when the seedlings were 21 days old) by adding salt (NaCl) gradually at the rate of 50 mM or less per day up to the specific level of each treatment. The gradual addition of salt was to avoid instantaneous stressing of the seedlings by high salinity levels. The boxes were divided into 8 groups of three boxes and each group has one of the salinity treatments as follows: 0, 5, 10, 20, 40, 80, 160 and 320 mM NaCl. The solutions were aerated constantly by compressed air.

Four harvests were taken, at 7-days intervals after salinization. Five plants were harvested then the plants were divided into two parts (shoot and root). Shoot and root length and leaf area were recorded. The fresh weight for shoot and root were recorded and then dried in an oven for 48 hours at 75°C. The oven dry weights were recorded and used to calculate the water content and relative growth rate (RGR) according to Pitman (1975): RGR= (loge w2- loge wl) / (T2-T1), where wl and w2 are weights at times Tl and T2.

RESULTS

The values of the growth parameters recorded in this work for *Calotropos porecera* seedlings were generally affected by high NaCl salinity concentrations. The shoot length increased with increasing plant age in all harvests except in the last two treatments (160 and 320 NaCl), at the same time their was a significant effect for increasing salinity treatments on the shoot length from 40 mM NaCl and up (Table 1). The root length increased with increasing salinity treatments and age when compared with control treatment except in the last two treatments (Table 2). The ileaf area increased with increasing salinity treatments in the second and third harvests but there was a significant decrease in the last four treatments (Table 3).

The shoot fresh and dry weights increased in the low salinity treatments (5 & 10 mM NaCl) and thin decreased with increasing salinity (Table 4 & 5), and their was a significant increase with increasing plant age except in the last two treatments. The root fresh weight gave the same trend of the shoot but there was no clear trend for the root dry weight (Table 6 & 7). The shoot water content show a constant amount in the shoot with both increasing age or NaCl salinity except in the last two treatments (160 & 320 mM Table 8). The root water content, however, there was no clear effect for salinity (Table 9).

The shoot and root relative growth rate (RGR) calculated in Table 10 and 11 promote as plant aged between the first three harvests in the low salinity treatments (0, 5, 10 and 20 mM NaCl), but it reduced with increasing salinity and age.

DISCUSSION

The growth of *Calotropis procera* measured over the short-term (three weeks) under different NaCl salinity concentrations give an indication that the seedlings of this plant differ in their response to NaCl salinity.

Growth stimulates at low salinity treatments (40 mM NaCl and less), it has been previously reported for other plants (*e.g.* Al-Zahrani, 1996). The plants not only grew under the low salinity concentration, but they continued to do so under higher salinities (from 160 mM NaCl and up), although to a limited extent. Perhaps because the inhibition effect of higher salinity concentrations in the metabolic process of plant (Bejaoui, 1985), or because of the effect in the cell elongation (Kurth, *et al.*, 1986) or because of that plant transferred it's energy from growth to accumulate suitable osmoregulation (Ownbey and Mahall, 1983).

Table (1) : Shoot length (cm) of *Calotropis procera* seedlings grown in different salinities (0, 5, 10, 20, 40, 80, 160 and 320 mM NaCl) for 28 days. (Mean ±SE).

Age (days)	Control	NaCl concentration (Mm)						
		5	10	20	40	80	160	320
28	6.3±0.9	6.2±0.5	6.5+0.5	6.3±0.3	6.2±0.2	6.0±0.2	5.5±0.3	5.4±0.5
35	11.8±0.2	11.8±0.9	10.7±0.4	10.1+0.6	10.3±0.8	8.0±0.5	5.2±0.2	4.8±0.1
42	14.0±1.1	15.3±0.3	15.1±0.6	13.1±0.9	12.2±0.8	8.9±0.6	5.2±0.7	4.5±0.8
49	20.3+1.6	20.7±0.6	20.7±1.0	20.4±1.0	15.6±0.7	11.2±1.3	5.2±0.6	4.2±0.5

LSD at 5%; salinity stress: 0.021; age: 0.014; interaction between stress and age: 0.125.

Table (2) : Root length (cm) of *Calotropis procera* seedlings grown in different salinities (0, 5, 10, 20, 40, 80, 160 and 320 mM NaCl) for 28 days. (Mean ±SE).

Age (days)	Control	NaCl concentration (mM)						
		5	10	20	40	80	160	320
28	9.4±0.7	9.2±2.6	13.3±1.4	10.6±0.6	10.3±0.9	10.0±1.0	6.7±1.2	4.7±1.1
35	21.2±0.7	22.5±1.7	22.1±2.6	19.4±3.4	18.8±3.0	14.5±3.8	6.6±0.3	4.5±0.1
42	26.9±3.5	29.3±1.0	34.5±0.7	32.8±3.5	29.2±4.0	28.1±1.4	6.9±1.3	4.2±0.1
49	33.6±1.8	38.1±2.2	40.9±7.0	42.0±4.1	41.9±3.1	38.4±2.6	6.3±0.8	4.3±0.6

LSD at 5%; salinity stress: 0.039; age: 0.029; interaction between stress and age: 0.276.

Table (3) : Leaf area (cm^2) of *Calotropis procera* seedlings grown in different salinities (0, 5, 10, 20, 40, 80, 160 and 320 mM NaCl) for 28 days. (Mean ±SE).

Age (days)	Control	NaCl concentration (mM)						
		5	10	20	40	80	160	320°
28	3.91±0.72	4.47±0.49	5.10±0.75	3.41±0.42	3.88±0.31	4.16±1.17	2.20±0.21	1.79±021
35	7.86±1.47	11.07±1.78	12.85±1.77	12.09±2.08	10.23±1.91	10.08±0.92	2.24±0.30	1.56±0.18
42	12.87±1.00	14.52±1.95	15.69±3.27	17.49±2.62	13.62±2.54	12.94±1.42	2.35±0.29	1.56±0.16
49	28.22±6.97	29.62±4.55	27.76±4.47	22.71±3.79	19.66±2.37	18.38±2.25	2.42±0.35	1.51±0.20

LSD at 5%; salinity stress: 0.086; age: 0.043; interaction between stress and age: 0.311.

Table (4) : Shoot fresh weight (g) of *Calotropis procera* seedlings grown in different salinities (0, 5, 10, 20, 40, 80, 160 and 320 mM NaCl) for 28 days. (Mean ±SE).

Age (days)	Control	NaCl concentration (mM)						
		5	10	20	40	80	160	320
28	0.18±0.01	0.17±0.01	0.15±0.01	0.16±0.01	0.14±0.01	0.12±0.01	0.07±0.01	0.04±0.01
35	0.25±0.01	0.28±0.01	0.37±0.01	0.39±0.02	0.25±0.01	0.24±0.01	0.02±0.002	0.01±0.001
42	0.59±0.02	0.74±0.03	1.10±0.07	0.86±0.04	0.42±0.03	0.38±0.05	0.02±0.002	0.01±0.001
49	1.38±0.32	1.49±0.5	1.62±0.09	1.13±0.09	0.66±0.03	0.41±0.05	0.02±0.003	0.01±0.002

LSD at 5%; salinity stress: 0.029; age: 0.019; interaction between stress and age: 0.102.

Table (5) : Shoot dry weight (g) of *Calotropis procera* seedlings grown in different salinities (0, 5, 10, 20, 40, 80, 160 and 320 mM NaCl) for 28 days. (Mean ±SE).

Age (days)	Control	NaCl concentration (mM)						
		5	10	20	40	80	160	320
28	0.015±0.001	0.010±0.000	0.010±0.00	0.011±0.001	0.010±0.00	0.0091±0.00	0.008±0.001	0.006±0.001
35	0.023±0.001	0.025±0.001	0.028±0.00	0.026±0.001	0.019±0.001	0.018±0.001	0.012±0.001	0.006±0.001
42	0.050±0.002	0.060±0.003	0.080±0.01	0.064±0.003	0.029±0.004	0.027±0.002	0.012±0.002	0.006±0.001
49	0.129±0.01	0.131±0.005	0.138±0.01	0.094±0.007	0.052±0.004	0.028±0.002	0.012±0.003	0.006±0.001

LSD at 5%; salinity stress: 0.048; age: 0.021; interaction between stress and age: 0.13

Table (6) : Root fresh weight (g) of *Calotropis procera* seedlings grown in different salinities (0, 5, 10, 20, 40, 80, 160 and 320 mMNaCl) for 28 days. (Mean±SE).

Age (days)	Control	NaCl concentration (mM)						
		5	10	20	40	80	160	320
28	0.08±0.01	0.07±0.00	0.07±0.01	0.08±0.01	0.08±0.01	0.07±0.01	0.03±0.00	0.004±0.001
35	0.11±0.01	0.12±0.01	0.13±0.01	0.16±0.01	0.13±0.01	0.11±0.01	0.02±0.01	0.004±0.001
42	0.18±0.03	0.19±0.04	0.21±0.03	0.22±0.02	0.21±0.03	0.17±0.03	0.02±0.01	0.003±0.000
49	0.23±0.07	0.30±0.01	0.33±0.05	0.41±0.05	0.31±0.09	0.20±0.03	0.02±0.01	0.003±0.002

LSD at 5%; salinity stress: 0.106; age: 0.059; interaction between stress and age: 0.402.

Table (7) : Root dry weight (g) of *Calotropis procera* seedlings grown in different salinities (0, 5, 10, 20, 40, 80, 160 and 320 mM NaCl) for 28 days. (Mean ±SE).

Age (days)	Control	NaCl concentration (mM)						
		5	10	20	40	80	160	320
28	0.007±0.001	0.006±0.00	0.007±0.00	0.008±0.001	0.008±0.001	0.008±0.001	0.004±0.001	0.001±0.0
35	0.01±0.001	0.01±0.001	0.01±0.001	0.01±0.001	0.02±0.001	0.02±0.003	0.004±0.001	0.001±0.0
42	0.03±0.006	0.03±0.01	0.03±0.009	0.03±0.007	0.03±0.003	0.03±0.005	0.006±0.001	0.001±0.0
49	0.06±0.01	0.08±0.005	0.08±0.01	0.06±0.01	0.06±0.01	0.04±0.009	0.006±0.004	0.001±0.0

LSD at 5%; salinity stress: 0.157; age: 0.044; interaction between stress and age: 0.490.

Table (8) : Shoot water content (%) of *Calotropis procera* seedlings grown in different salinities (0, 5, 10, 20, 40, 80, 160 and 320 mM NaCl) for 28 days. (Mean ±SE)

Age (days)	Control	NaCI concentration (mM)						
		5	10	20	40	80	160	320
28	91.7±1.5	94.1±3.9	933±1.0	93.1±1.5	92.9±2.3	92.5±0.0	88.6±3.9	85.0±6.8
35	90.8±1.9	91.1±0.9	92.4±1.5	93.3±0.9	92.4±2.0	92.5±1.9	40.0±5.1	40.0±1.9
42	91.5±2.1	91.9±2.1	92.8±1.6	92.6±2.0	93.1±2.7	92.9±1.5	40.0±5.0	40.0±3.0
49	90.7±2.2	91.2±1.1	91.5±2.7	91.7±2.8	92.1±2.7	93.2±1.8	40.0±13.0	40.0±1.3

LSD at 5%; salinity stress: 0.495; age: 1.229; interaction between stress and age: 0.552.

Table (9) : Root water content (%) of *Calotropis procera* seedlings grown in different salinities (0, 5, 10, 20, 40, 80, 160 and 320 mM NaCl) for 28 days. (Mean ±SE).

Age (days)	Control	NaCl concentration (mM)						
		5	10	20	40	80	160	320
28	91.3±1.0	91.4±1.1	90.0±1.1	90.0±1.7	90.0±0.5	88.6±1.2	86.7±1.6	75.0±3.2
35	90.9±1.0	91.7±1.7	92.3±1.4	93.8±1.0	84.7±0.6	81.8±1.8	80.0±13.2	75.0±4.9
42	83.3±1.5	84.2±1.8	85.7±2.8	90.9±2.4	86.2±0.9	76.5±0.2	70,0±5.4	66.7±4.1
49	73.8±1.0	73.4±1.6	75.7±0.3	85.4±1.4	80.9±0.5	75.0±0.8	70.0±7.0	66.7±3.3

LSD at 5%; salinity stress: 0.063; age: 0.051; interaction between stress and age: 0.307.

Table (10) : Shoot relative growth rate (RGR, g/g/day) of *Calotropis procera* seedlings grown in different salinities (0, 5, 10, 20, 40, 80, 160 and 320 mM NaCl) for 28 days. (Mean ±SE).

Age (days)	Control	NaCl concentration (mM)						
		5	10	20	40	80	160	320
28-35	0.061	0.131	0.147	0.123	0.092	0.099	0.058	0.00
35-42	0.111	0.125	0.150	0.129	0.061	0.058	0.000	0.00
42-49	0.135	0.112	0.078	0.055	0.084	0.005	0.000	0.00

Table (11) : Root relative growth rate (RGR, g/g/day) of *Calotropis procera* seedlings grown in different salinities (0, 5, 10, 20, 40, 80, 160 and 320 mM NaCl) for 28 days. (Mean ±SE).

Age (days)	Control	NaCl concentration (mM)						
		5	10	20	40	80	160	320
28-35	0.051	0.073	0.051	0.032	0.131	0.131	0.00	0.0
35-42	0.157	0.157	0.157	0.157	0.058	0.099	0.099	0.0
42-49	0.099	10.140	0.140	0.099	0.099	0.032	0.000	0.0

Dry matter production of seedlings increased by salinity up to 10 mM NaCI, moreover, 5 and 10 mM NaCI stimulate a pronounced increase in seedling growth (shoot length, fresh and dry weights and root dry weight and leaf area). Different workers (Heikal, *et al,* 1982; Shaddad and Zidan, 1989 and Zidan and Al-Zahrani, 1994) recorded a similar results for some other plants. The calculated relative growth rate (RGR) give clear indication that the seedlings continued growth with increasing both salinity and age except in the last two treatments (160 and 320 NaCI). RGR generally declined as plant aged and since growth is depressed with increasing salinity concentration, RGR most be lower (Al-Zahrani, 1990).

The higher water content in the shoot and root of *Calotropis procera* seedlings of alinized plants was in agreement with the results obtained by other workers (Zidan, 1979 and Zidan & Al-Zahrani, 1994).

Finally, the growth of *Calotropis procera* seedlings was greatly reduced by high salinity toncentration, which indicated the sensitivity of this plant to high NaCI salinity and can tolerate low and medium salinity. More investigation need to be done for some metabolic processes.

REFERENCES

1. Al-Zahrani, H.S. (1990). Effect of environmental factors on the physiology of halophyte *Suaeda maritima* (L) Dum. D.Phil. Thesis, University of Sussex; p. 191.
2. Al-Zahrani, H.S. (1996). Responses of *Ocimum basilicum* L. (Sweet Basil) to edaphic stress. (1) Effect of salinity on germination, photosynthetic pigments and growth. *Bulletin of Pure and Applied Sciences,* 15B;p. 13-21.
3. Bejaoui, M. (1985). Interaction between NaCI and some phytohormones on soybean growth. *Plant Physiol,* 120; p. 95-110.
4. Flowers, T.J.; Flowers, S.A. and Greenway, H. (1986). Effect of sodium chloride on tobacco plants. Plant, *Cell and Environment,* 9; p. 645-651.
5. Hoagland, D.R. and Anion, D.I. (1950). The water culture method for growing plants without soil. Calif. Agric. Exp. Sta. Cir.; p. 347-372.
6. Heikal, M.; Shaddad, M. and Ahmed, A. (1982). Effect of water stress and gibberellic acid on germination of flax, sesame and onion seeds. *Biologia Plantarum (Praha),* 24; p. 124-129.
7. Kurth, E; Cramer, G; Lauchli, A. and Epstein, E. (1986). Effect of NaCI and $CaCl_2$ on cell en-largement and cell production in cotton roots. *Plant Physiol,* 82; p. 1102-1106.
8. Larcher, W. (1995). Physiological plant ecology. 3rd Ed. Springer, New York.
9. Migahid, A.M. (1990). Flora of Saudi Arabia. 4th Ed. Riyadh. Saudi Arabia.
10. Ownbey, R. and Mahall, B. (1983). Salinity and root conductivity differential responses of a coastal succulent halophyte *Salicornia virginica* and a weedy glycophyte; p. 1016.
11. Pitman, M.G. (1975). Whole plants. In: ion transport in plant cells and tissues. Ed. By D. Baker and J. Hall. North-Holland Publishing Company.
12. Shaddad, M.A. and Zidan, M.A. (1989). Effect of NaCl salinity on the rate of germination, seedling growth, and some metabolic changes in *Raphanus satvus* L. and *Trigonila Foenum-graecum* L. Beitr. Trop. Landwirtch. *Vet. Med.* 27; p. 187-194.
13. Zidan, M.A. (1979). Photosynthetic Activity, mineral composition and growth of some legumes as affected by salinity. MSc. Thesis, Assiut Uni. Egypt; p. 150.
14. Zidan, M.A. and Al-Zahrani, H.S. (1994). Effect of NaCl on the germination, seedling and some metabolic changes in sweet basil *(Ocimum basilicum). Pak. J. Sci. Ind. Res.* 12; p. 541-543

Chapter 28

Enhancing the Sustainability of Quinoa Production and Soil Resilience by Using Bioproducts Made with Native Microorganisms

Noel Ortuño*, José Antonio Castillo*, Mayra Claros, Oscar Navia, Marlene Angulo, Daniel Barja, Claudia Gutiérrez and Violeta Angulo

Fundación PROINPA, Av. Meneces Km 4, El Paso, Cochabamba, Bolivia; E-Mails: m.claros@proinpa.org (M.C.); o.navia@proinpa.org (O.N.); m.angulo@proinpa.org (M.A.); d.barja@proinpa.org (D.B.); c.gutierrez@proinpa.org (C.G.); v.angulo@proinpa.org (V.A.)

* Authors to whom correspondence should be addressed; E-Mails: n.ortuno@proinpa.org (N.O.); j.castillo@proinpa.org (J.A.C.); Tel.: +591-4-431-9595; Fax: +591-4-431-9600.

ABSTRACT

Microorganisms are involved in a network of interactions with plants, promoting growth and acting as biocontrol agents against diseases. In this work, we studied native microorganisms associated with quinoa plants (*Chenopodium quinoa*) and the application of these organisms to the organic production of quinoa in the Andean Altiplano. Quinoa is a non-cereal grain native to the Andean highlands and is highly nutritious and gluten-free. As such, the international demand for quinoa has increased substantially in recent years. We isolated native endophytic bacteria that are able to fix nitrogen, solubilize phosphate and synthesize a phyto-hormone and native strains of *Trichoderma*, a fungus typically used for increasing plant growth and tolerance to biotic and abiotic stresses. Greenhouse assays and field trials allowed for selecting promissory bacterial isolates, mostly belonging to *Bacillus* and *Paenibacillus* genera, that increased plant length, panicle weight and grain yield. Selected microbial isolates were large-scale multiplied in simple and inexpensive culture media and then formulated to obtain bioproducts that were distributed among local farmers. Thus, we developed a technology for the exploitation of beneficial microbes, offering promising and environmentally friendly strategies for the organic production of quinoa without perturbing the native microbial diversity of Andean soils and making them more resilient to the adverse effects of climatic change and the over-production of quinoa.

Keywords

Plant growth-promoting bacteria; *Trichoderma*; endophytes; quinoa; Andean Altiplano; soil

1. INTRODUCTION

Soil is a fundamental and irreplaceable natural resource, providing the essential connection between the land, air and water resources that allow life on Earth. As an example, it maintains the life of microorganisms that are involved in bio-geochemical cycles, nutrient recycling and carbon storage. Soil also sustains agricultural production, which, together with the oceans, is the most important source of food for animals and humans. However, soil resources are threatened due to degradation and desertification. According to the Food and Agriculture Organization [1], soil degradation is a "process which lowers the current and/or potential capability of soil to produce goods and services". Thus, soil degradation is a threat that national and international organizations are trying to hinder and reverse.

The flat land area located between the Andes Mountains, known as the Altiplano, shared between Peru and Bolivia, is facing a rapid process of soil erosion, due to increasing drought, climate change, population growth and altered farming practices. This process leads to desertification, which results in soils with low vegetable cover and very poor organic matter content. These problems have caused drastic decreases in the production of crops, reducing the already low incomes of the farmers. In some cases, the farmers have used agrochemicals to mitigate soil fertility problems, with the potential of negative effects on the environment. In parallel, there is an increasing demand in international markets for organic products, such as the quinoa (*Chenopodium quinoa*). Quinoa is a native pseudograin from the Altiplano, where local people domesticated it since ancestral times. Despite its ancestral condition, quinoa has lately become particular popular, due to its gluten-free and high nutritional value properties (with an average of 14.8% protein and an exceptional balance between oil, protein and starch, [2]). During the last 10 years, Bolivian quinoa exportation increased 26 times in value and nine times in volume [3]. This is the reason why quinoa, the most economically attractive crop from this area, is urging Andean growers to intensify its production beyond sustainability.

Organic production is a national priority in countries like Bolivia and has been mainly driven by the international demand for organic products for exportation.

This motivates the agricultural sector, which seeks to produce food respecting the environment and the health of farmers. For this reason, it is necessary to develop adequate technologies to support organic production and according to local requirements. These technologies include bioproducts based on living microorganisms that are used as biofertilizers and biocontrol agents for pests and diseases, depending on the microbial species with which they have been made. Bioproducts are important tools to help reduce the harmful effects of chemical fertilizers and pesticides in the environment and to produce food free of contaminants and at a lower cost, in this way helping the economy of small farmers. Bioproducts can also help to improve the quality of the soils in the Altiplano by retaining microbial populations, which will help to mitigate soil degradation and enable farmers to produce organic crops, including quinoa, broad beans and potato, among others).

In this work, we focused on the isolation of native microorganisms associated with quinoa at the Bolivian Altiplano and studied their nutrient recycling potential and beneficial effects on plant growth. The best microbial isolates were used to formulate bioproducts for organic production, which were distributed among local farmers. We intentionally sought to work with native microorganisms to reduce the disturbance of the Altiplano soil ecosystem and to keep the technology as simple as possible, in order to provide cost-effective technologies for Andean farmers.

2. RESULTS AND DISCUSSION

2.1. Microbial Isolates

Given the need for ecological fertilizers and biopesticides for organic agriculture, our investigation has been directed towards the development of new biotechnologies, to contribute to the clean and healthy production of quinoa and to help mitigate soil degradation. In the last few years, there has been a growing interest in plant growth-promoting bacteria (PGPBs) and fungi (PGPFs), because they can recycle nutrients, promote plant growth and, in some cases, prevent the infection of plant tissue by pathogens. Such microorganisms can be symbiotic or free living [4,5]. To reach this goal, we explored the native microorganisms associated with quinoa plants, isolating and identifying a series of bacterial groups. Results indicate that one hundred and four different bacterial isolates were obtained from internal plant tissue (endophytic bacteria), of which fifty five are from roots, eight from stems and forty one from leaves. Of the fungi group, only rhizospheric individuals were isolated. Bacteria and fungi were molecularly identified (see below), consisting of the following:

Rhizosphere microorganisms: Fungi individuals were isolated that belong to the following genera or species: *Trichoderma harzianum*, *T. asperellum*, *T. koningiopsis*, *Beauveria bassiana*, *B. brongniartii* and *Metarhizium* sp.

Endophytic microorganisms: Bacterial species isolated from leaves: *Bacillus amyloliquefaciens*, *B. tequilensis*, *B. subtilis*, *B. pumilus*, *B. licheniformis*, *B. horikoshii*,

B. atrophaeus, *B. thuringiensis* and *Paenibacillus* sp. Bacterial species isolated from stems and roots: *B. aryabhattai*, *B. subtilis*, *B. horikoshii*, *B. megaterium*, *Pseudomonas* sp., *Paenibacillus* sp. and *P. odorifer*.

2.2. Functional Activities of Isolated Microorganisms Promoting Plant Nutrition and Growth

Functional tests performed under *in vitro* conditions allowed us to analyze the bacterial isolates that we obtained from quinoa plants for the evidence of nitrogen fixation, phosphate solubilization and plant growth hormone (indole-3-acetic acid, IAA) production capabilities. From all endosphere isolates, a number of them showed desired features: P-solubilization (72 isolates), N-fixation (30 isolates) and IAA production (59 isolates). Some isolates exhibited multiple functions. Thus, almost 70% of bacterial isolates exhibited beneficial activity, which is consistent with other works that aimed to obtain bacteria with the three tested properties [6,7]. These bacterial isolates belong to bacterial groups well known as PGPBs (*i.e.*, *Bacillus*, *Paenibacillus* and *Pseudomonas* [8,9]).

Although we identified a range of fungi species, we only focused on isolating and characterizing individuals from the *Trichoderma* genus, since members of this group are well recognized as PGPF and as biocontrol agents against a wide range of soil-borne pathogens [10,11].

The conservation of the isolates was structured under laboratory conditions, considering origin, date of harvesting, altitude and plant tissue information. Storing conditions were 4 °C, a relative humidity of 50% and tryptone soy agar (TSA) as the culture medium. This collection constitutes a valuable source of bacteria and fungi strains for future research on agricultural and biotechnological applications.

2.3. Greenhouse Testing

Bacterial isolates that showed positive results for the three tests performed under *in vitro* conditions were used to inoculate plants under greenhouse conditions in order to confirm their beneficial properties in an *in vivo* assessment. After a post-inoculation period of eight weeks, quinoa plants were harvested and measured for plant height, panicle weight and root weight, volume and length. Statistical analysis showed that the treatments with bacterial isolates did not influence the root weight, volume or length, but increased plant height and panicle weight (Figure 1). The bacterial isolates that increased plant height were 1Bp and 5Bp (*B. pumilus*, P-solubilization bacteria), 139 and 143 (*B. simplex*, N-fixing bacteria), 2p, 3p and 4p (*Paenibacillus* sp., N-fixing bacteria) and BAQ-11 (*B. subtilis*, N-fixing bacteria) and for panicle weight were 3p and 4p (*Paenibacillus* sp.) and BAQ-11 (*B. subtilis*) (Figure 1). Thus, some bacterial isolates that increased panicle weight also increased plant height. All other isolates showed the same behavior on plant height and panicle weight compared to the uninoculated control.

2.4. Selection of the Culture Media of the Substrates for the Large-Scale Multiplication of the Microorganisms

Selected microbial isolates were evaluated for large-scale growth in different media and field-tested by local farmers. We sought to develop a simple, but efficient and, at the same time, inexpensive technology for the multiplication of microorganisms and the formulation of bioproducts; hence, farmers could use this technology for the preparation of their own bioproducts. A culture media for large-scale multiplication of bacteria mostly from the *Bacillus* genus was developed. For this, six different liquid media were tested (see Materials and Methods).

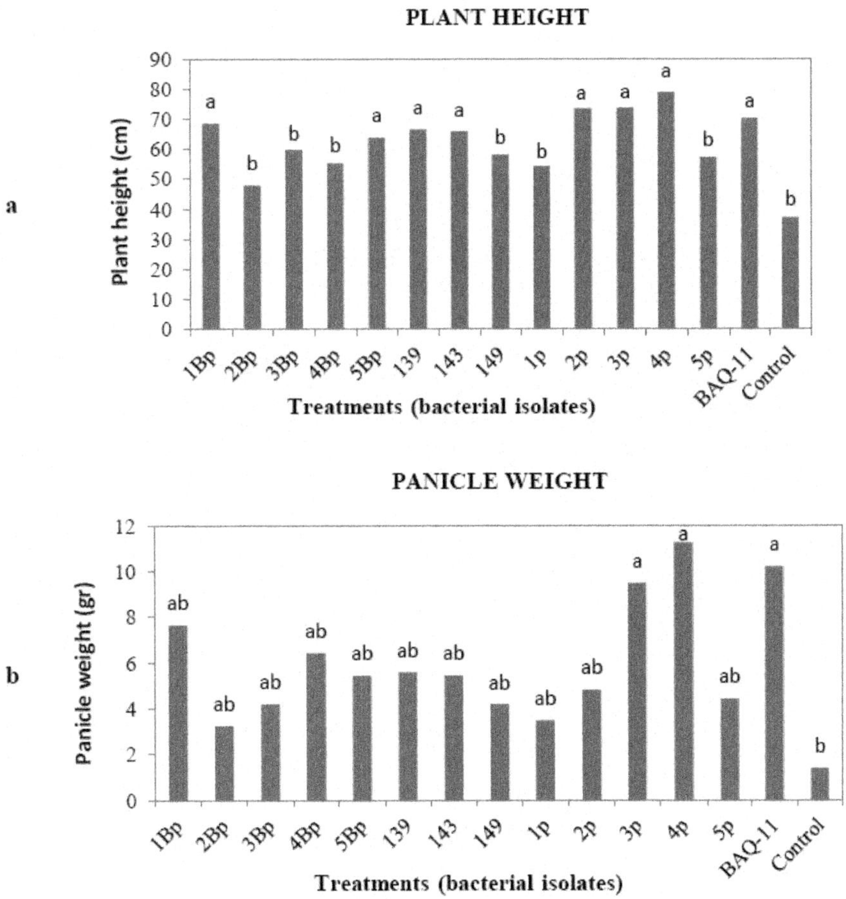

Figure 1. The results of **(a)** plant height and **(b)** panicle weight of eight-week-old quinoa plants treated with bacterial suspensions (10^8 cfu/mL) of different bacterial isolates are shown; control plants represent un-inoculated plants. Height and weight are measured in centimeters and grams, respectively; bacterial isolate identity: 1Bp, 2Bp, 3Bp, 4Bp and 5Bp (*B. pumilus*); 139, 143 and 149 (*B. simplex*); 1p, 2p, 3p, 4p and 5p (*Paenibacillus* sp.); BAQ-11 (*B. subtilis*).

The results, detailed in Figure 2a, indicate that culture medium elaborated with soy protein is statistically superior to the other treatments; consequently, this medium was employed for further large-scale multiplication of *Bacilli* bacteria. Once the means of culture and the aerobic system were developed, a bioreactor for the multiplication of the bacteria to a greater scale was built. A stainless steel bioreactor (Figure 3a) allowed a production in volumes that enabled a supply for as much as 500 hectares of quinoa in the Andean Altiplano. However, this production is still insufficient, since the extension of quinoa land in the Altiplano is much larger.

Similar to the bacteria production, we sought to identify a suitable and inexpensive substrate for large-scale *Trichoderma* production (*T. harzianum* and *T. koningiopsis*). We tested seven different substrates for *T. harzianum* growth and spore development. There were significant differences among substrates, with substrate 1 (rice grain), followed by substrate 5 (rice grain:husk, 1:1) and 6 (rice grain:husk 1:2.33), showing the most potential (Figure 2b). Fungi spores were used for the formulation of bioproducts by mixing them with inert material.

2.5. Formulation of the Bacteria in Powder

For bioproduct formulation in powder form, the bacteria were mixed with inert media ($CaCO_3$) and placed on metal trays to be dried (Figure 3b). The material was then ground into a fine powder and evaluated for bacteria re-growth and pH and biological property maintenance. Results showed that bacteria from bioproducts were able to grow in culture media, as well as the non-formulated bacteria [12]. To improve the efficiency of producing the bioproducts, a drying apparatus with a capacity to process 350 kg per 36 hours was designed and constructed (Figure 3c). Using this equipment, we anticipate that sufficient bioproduct for 500 hectares can be produced per month, in this way giving the opportunity to supply bioproducts to a larger number of small farmers. The bioproducts (based on bacteria or fungi) were packed in plastic bags (Figure 3d) and named with striking names for distribution among farmers (Figure 3e).

a

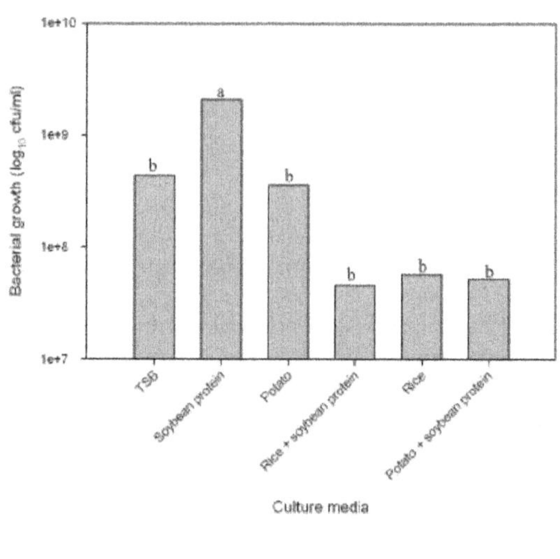

b

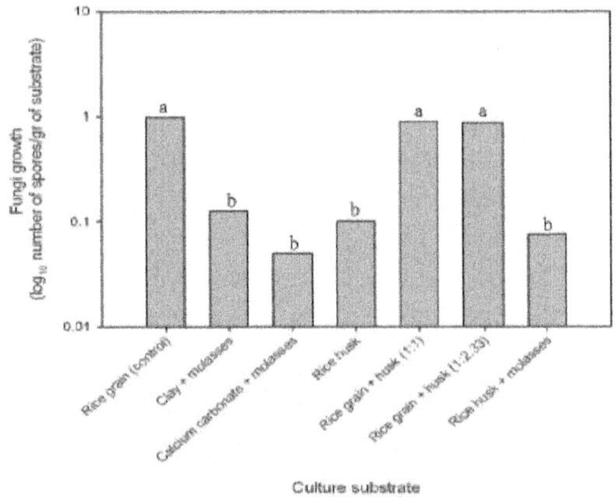

Figure 2. (**a**) *Bacillus* and (**b**) *Trichoderma* isolates grown in different culture media and substrates, respectively tested to define the best means for a large-scale and inexpensive multiplication of microorganisms. Bacteria or fungal growth was measured calculating the number of colony forming units or spores per milliliter or gram, respectively. TSB, tryptone soy broth.

Figure 3. (**a**) Stainless steel bioreactors designed for a massive production of *Bacillus subtilis* spores; (**b**) *Bacillus subtilis* culture broth after mixing with inert material (aluminum/calcium silicate); (**c**) Dryer with forced air for the powder formulation of bacteria; (**d**) Packing of bioproducts. (**e**) Biofertilizers ready for distribution among farmers for organic production; (**f**) Organic quinoa production in the Southern Altiplano (highlands) of Bolivia: at the left quinoa, field supplemented with a bioproduct (Tricobal: *B. subtilis, T. koningiopsis and T. harzianum*); at the right, control quinoa field without the use of any bioproducts.

2.6. Field Trials: Response of Quinoa to Treatments with Bioproducts

It is important to screen the response of quinoa plants to microbial inoculation under field conditions since *in vitro* screenings for PGPBs do not always reflect the reality of production on the field *(i.e.,* microbial performance *in situ* is different from *in vitro*). Thus, the bioproducts were tested under organic production of quinoa by local Altiplano farmers for plant growth performance and disease control responses. The bioproducts, called Biobacillus (composed of *B. subtilis* strain BAQ-11) and Tricotop (containing *T. koningiopsis* y *T. harzianum*), were tested.

Table 1 shows the results indicating that Biobacillus or Tricotop increased quinoa yield compared to the control without microorganism amendment. Interestingly, using both products, the highest yields were obtained. This led to the idea of creating a third bioproduct, called Tricobal, which is formulated with *B. subtilis, T. koningiopsis* and *T. harzianum*.

Table 1. Comparison of the average yield (kg/ha) of quinoa with the use of bioproducts.

Treatment	Average yield (kg/ha) *
Control (no bioproduct)	1,051.67 c
Biobacillus	1,206.67 b
Tricotop	1,346.67 b
Tricotop + Biobacillus	1,516.67 a

* Different letters mean statistical differences between treatments of $p = 0.05$.

We evaluated Tricobal over two quinoa production seasons (two different years). Tricobal was soil-applied to farmers' fields at the time of planting. We measured the plant growth, leaf area, panicle size and average grain yields. Table 2 shows that the differences between treatments with bioproduct (Tricobal) and the controls were significant. Plants treated with Tricobal looked overall healthier and more vigorous than untreated plants (Figure 3f). This result demonstrates that Tricobal is a potential tool for farmers to improve the crop productivity of quinoa.

Table 2. Bioproduct amendment effect on the development and yield of quinoa.

Year	Treatment	Plant size (cm) ^{m}	Leaf area (%) ^{m}	Panicle length (cm) ^{m}	Yield (qq/ha) ^{m,n}
1 (2009)	Tricobal	125 a	70 a	40 a	20 a
	Control (no bioproduct)	80 b	50 b	25 b	11 b
2 (2010)	Tricobal	150 a	80 a	2 a	25 a
	Control (no bioproduct)	110 b	55 b	30 b	12 b

^{m} Different letters mean statistical differences between treatments of $p = 0.05$; ^{n} qq: quintals (46 kg) per hectare.

3. EXPERIMENTAL SECTION

3.1. Microorganisms Isolation

To explore the microbe biodiversity of the Altiplano soil associated with quinoa plants, twenty organic quinoa crop fields were sampled from each Altiplano region: southern (S 19°41′0″, W 66°34′0″), central (S 18°3′0″, W 66°59′0″) and northern (S 17°4′0″, W 68°9′0″), close to Uyuni, Quillacas, Salinas de Garci Mendoza, Jalsuri, Lacaya and Cohana villages, which corresponds to the districts of Potosí, Oruro and La Paz, respectively. Soil samples were obtained by discarding the top 5 cm and collecting the next 20 cm. Plant root samples were taken 15 cm from the tip, stems from the section above the roots and leaves from the upper segment of the plant. All samples were transported to the laboratory under cold temperature for microorganism isolation.

Microorganisms were isolated on culture plates using specific culture media to separate bacteria (TSA) and fungi (potato dextrose agar (PDA)). For rhizosphere microorganism isolation, 10 g of roots with soil were washed with 90 mL of 0.85% NaCl and gentle agitation. Soil residues were vigorously shaken and then plated on respective medium, making dilutions until 10^{-9}. In order to isolate endophytic bacteria, small pieces of leaves (about 1 cm^2), stems and roots (1 cm-long) were cut and submerged in 15 mL of potassium phosphate buffer (0.1 M, pH 7.0) for 3 h with gently shaking. After this step, plant samples were directly treated with 70% ethanol for 1 min and then with 1.2% sodium hypochlorite for 15 min for superficial sterilization. After rinsing (four times with sterile 0.85% NaCl solution), the tissue fragments were smashed in 30 mL of sterile 0.85% NaCl solution. Serial dilutions were spread on TSA medium and incubated at 28 °C for 72 h. To isolate only endospore-forming bacteria, dilutions were subject to heat (75 °C for 15 min) before plating.

Bacterial colonies were studied at the morphological (color, shape, growing speed, *etc.*), biochemical (catalase, potassium hydroxide and oxidase) and cellular (Gram stain, flagella and spore presence/absence) level. Taxonomic identification of each bacterial and fungal isolate was performed by following the method described below.

3.2. Establishing the Functional Activity of Microbial Isolates

We attempted to determine the nitrogen (N) fixing, phosphorus (P) solubilizing and growth-promoting activities of bacterial isolates by means of the following methods:

To select bacteria able to fix atmospheric nitrogen, the leaves and roots were tightly washed and then treated with ethanol 70%. Five rinses were done with sterile distilled water to remove ethanol. Under sterile conditions, small fragments (0.5 cm) were cut and deposited on N-free Rennie media (2 g/L agar). Plant tissue fragments were incubated for 6 days at 30 °C without shaking. After incubation, a veil and then a film were grown, which constitutes an inoculum that was later diluted and plated on the same and fresh medium. After an incubation period of 48 h at 30 °C, isolated colonies were obtained. Since the growth medium contains tiny amounts of nitrogen, all bacteria that were able to grow on it could be nitrogen fixers. To confirm this supposition, we determined the identity of bacteria by means of the molecular techniques detailed below.

To select phosphorus-solubilizing bacteria, isolates were culture on NBRIP (National Botanical Research Institute's phosphate) medium [13] at 37 °C for 2 days. After this period, a clear halo was recorded as positive, which means that the phosphate was solubilized.

To select growth-promoting bacteria, we focused on finding isolates that produce IAA. Bacteria were cultured in tryptone soy broth (TSB) supplemented with 5 mM of L-tryptophan at 28 °C for 7 days. After the incubation period, cultures were centrifuged (3802× *g* for 5 min), and an aliquot of 75 μL of the supernatant was mixed with 225 μL of Salkowski reagent for 30 min under dark conditions. Reddish colored supernatant was read as a positive signal [14].

3.3. Taxonomic Identification of Isolates

To determine or confirm the taxonomic affiliation of the selected microorganisms, we extracted genomic DNA from each isolate following the protocol described by Wilson [15], for bacteria, and Melo and collaborators [16], for fungi. We have amplified by PCR a small genomic region that is conserved among all bacteria using universal primers (Table 3), and then, the amplified fragments were sequenced. For PCR, Phusion high fidelity DNA polymerase (manufactured by NEB, Ipswich, MA, USA) and the following conditions were used: each 25-μL amplification reaction mixture comprised ~10 ng chromosomal DNA, 1× HF (high-fidelity) buffer, 0.4 μM forward and reverse primers, 0.2 mM deoxynucleotide triphosphates (NEB), 1× Q reagent (Qiagen, Germantown, MD, USA) and 0.5 U Phusion DNA Polymerase. Reaction conditions for all the primers were as follows: initial denaturation at 95 °C for 10 min, 35 cycles of denaturation at 95 °C for 40 s, primer annealing at 55 °C for 40 s for bacteria and 52–59 °C for 1 min for fungi and extension at 72 °C for 1 min, followed by a final extension step of 72 °C for 5 min. The amplification product was then purified using Qiaquick columns (Qiagen) following the manufacturer's protocol before being used in a sequencing reaction. DNA sequencing was performed at the DNA sequencing facility, the University of Chicago, Chicago, IL, USA, following the standard Sanger sequencing protocol. Resulting sequences were edited and trimmed with the BioEdit program [17] and then used to find the most similar sequences in the database using the BLAST (Basic Local Alignment Search Tool) program [18]. Only matches reaching 99%–100% of identity and covering most of the sequence (>90%) were considered to assign a taxonomic name to the isolates. For analysis of *Trichoderma* sequences, TrichoBLAST and TrichOKEY programs [19] were also used to analyze DNA sequences from intergenic regions of ribosomal genes and internal segments of the *tef1* gene [20,21].

Table 3. Primers used in this work for the identification of microbial isolates; the same primers were used for amplification and sequencing.

Microorganism	Primer name	Gene marker [a]	Reference
Bacteria	27F	SSU rRNA	[22]
	1492R	SSU rRNA	[22]
Trichoderma	ITS1F	SSU rRNA	[23]
	ITS5	SSU rRNA	[24]
	ITS4	LSU rRNA	[24]
	LR1	LSU rRNA	[24]
	SR6R	SSU rRNA	[24]
	EF1-728F	*tef1* gene	[25]
	TEF1LLErev	*tef1* gene	[26]

[a] SSU rRNA, small subunit ribosomal RNA gene; LSU rRNA, large subunit ribosomal RNA gene.

3.4. Plant Inoculation with Beneficial Bacteria under Greenhouse Conditions

Bacterial isolates that showed functional activity on N-fixing, P-solubilizing and IAA production, as assayed in culture plates, were selected for further analyses on their performance *in planta*. Each pot containing 1 kg of sterile soil (silt, sand, organic matter mixed in equal parts) was inoculated with 0.5 mL of a bacterial suspension (1×10^8 cfu/mL). Then, three quinoa seeds were sown per pot.

After two months, plants were harvested, and the following parameters were measured: plant height (from the neck to the tip of the panicle, in centimeters); panicle weight (in grams); root weight (discarding neck portion, in grams); root volume (calculating the difference in volume of water displaced in a beaker); and root length of each root from bottom to tip (in centimeters). The assay was intended to follow a complete randomized blocks design, and data were analyzed by the Duncan applied test ($p = 0.05$) with four repetitions per treatment.

3.5. Large-Scale Multiplication of Bacterial Isolates

For bacteria-based bioproducts, a starter inoculum was prepared by multiplying *B. subtilis* strain BAQ-11 in TSB medium and incubating them for 3 days at room temperature with agitation until reaching a bacterial concentration of 1×10^8 cfu/mL. Different liquid culture media were tested to define which media was able to produce the highest number of colony forming units per milliliter. We used five different broths made of: (1) tryptic soy broth (commercial culture medium, control); (2) soybean protein (a subproduct from the edible oil industry); (3) potato (*Solanum tuberosum* L. ssp. *andigenum* cv. Waych'a); (4) a mix of rice grains and soybean protein (1:1 proportion); (5) rice grains; and (6) a mix of potato and soybean protein (6:1 proportion). These broths were prepared by boiling the product in water for 1 h and then were autoclaved for sterilization. The inoculated broths were incubated in a shaker at 100 rpm and 35 °C for ten days. After incubation, staining was performed with malachite green to verify bacterial spore formation [27]. To determine the number of spores/milliliter, three dilutions were plated on PDA medium for each treatment, using three repetitions of each dilution and 20 observations per repetition. Analysis of variance with four repetitions was employed to statistically validate the differences among treatments.

The best culture medium as assayed above was used for large-scale multiplication. Inoculum grown in TSB medium was employed to establish a larger culture (80 L). Forced sterile air was supplied without agitation, since air itself generates turbulence in the liquid. The culture was incubated at 22–25 °C for 10 days to reach a cell density of 1×10^9 spores/mL. This culture was used to formulate bioproducts (see below).

3.6. Large-Scale Multiplication of Fungi Isolates

The starter inoculum of *Trichoderma* species was obtained by growing single spores of the fungus on PDA medium. To determine the best substrate for large-scale and fast multiplication of *T. harzianum*, the starter inoculum was mixed with different sterile substrates: (1) rice grain; (2) clay + sugarcane molasses; (3) calcium carbonate + sugarcane molasses; (4) rice husk; (5) rice grain + rice husk (proportion 1:1); (6) rice grain + rice husk (proportion 1:2.33); and (7) rice husk + sugarcane molasses. Each substrate was incubated for 9 days at 28 °C, and at the end of this period, the number of spores per gram of substrate was measured using the Neubauer chamber under microscope as evidence for fungus growth. Spore calculation was obtained by adding the number of spores of five small squares (0.0025 mm^2) and multiplying it by 5000. A variance analysis applied to four repetitions of each treatment helped to statistically define the best substrate for the fungus growth.

The best culture medium as assayed according to the previous experiment was used for large-scale multiplication. The PDA-medium-based starter inoculum was employed to establish a larger culture in sterile rice grains and incubated for 15 days at 28 °C. In this way, a larger fungi biomass was achieved, which, in turn, was employed to inoculate the selected substrate (rice mixed with rice husk (1:1)) for large-scale multiplication. After a period of 15 days, the multiplication medium containing *Trichoderma* spores was dried on trays, which stresses the fungus, generating more spores. Spores were separated from the multiplication medium by sieving and then were used to formulate bioproducts, as described below.

3.7. Formulation of Bioproducts

For the formulation of bioproducts under powder form, we mixed bacterial broth or *Trichoderma* spores with bentonite (aluminum/calcium silicate) in a relation of 1 L of bacterial broth or 100 g *Trichoderma* spores to 1 kg of bentonite. Subsequently, the mix containing the bacterial broth was placed in plastic trays and dried at room temperature using forced air stream for three days to obtain a completely dry product. The product was ground using a hammer mill to obtain a fine powder, which was directly bagged or resuspended in water and dispensed into bottles for the liquid design. The viability of bacterial cells was analyzed to assure a product containing living microorganism ($\sim 1 \times 10^9$ cfu/mL).

3.8. Plant Inoculation with Bioproducts under Open Field Conditions

At the time of planting (late November), about 7 kg/ha of decomposed sheep manure were deposited as the basal dressing, and then, four quinoa seeds from the "Real Blanca" variety were sown per spot on Altiplano soil, which is mostly sandy (Quillacas area, Oruro district). Bioproducts (Tricotop, Biobacillus or Tricobal) were applied to the soil at doses of 1 kg/ha each. Later and at the branching stage, Biobacillus was also used at 1 kg/ha as foliage fertilizer for all assays, except the control. After two months, quinoa plants were harvested, and the response variables evaluated were plant size, leaf area, panicle length and yield. We assayed the bioproducts on 20 ha (40,000 plants/ha) and evaluated another 20 ha as the control without bioproducts. For the statistical analysis, we used a completely random block design with a split plot layout, where the main plot was the organic matter and small plots were: control (no bioproduct), Biobacillus, Tricotop, Biobacillus + Tricotop or Tricobal. Then, the means were compared by the Duncan applied test ($p = 0.05$).

4. CONCLUSIONS

Plant growth-promoting bacteria and fungi (*Trichoderma*) were isolated from quinoa tissues and evaluated for their individual and combined effects on the growth promotion of quinoa and the increased production of grain. To our knowledge, this is the first report that deals with the analysis of microorganisms associated with quinoa that grow in its native habitat, the Andean Altiplano. After the large-scale multiplication and formulation steps, we obtained bioproducts by a simple and economical means, which will allows for an environmental friendly production of organic quinoa for international markets, at the same time keeping the native populations of microorganisms in the soil. Undoubtedly, the microorganisms isolated and studied in this work constitute a valuable contribution to the organic production of quinoa (and other crops) under the extreme environmental conditions of the Andean Altiplano and could help to substantially improve the soil resilience of the small-holder quinoa farming systems in the Andean Altiplano.

Acknowledgments

This work was supported by the "Fondo Regional de Tecnología Agropecuaria" (FONTAGRO), grant number 7075 to Fundación Proinpa and "Consorcio Andino", funded by the Netherlands Directorate-General for International Cooperation. We wish to thank Robert Gallagher for technical advising and language corrections.

Author contributions: Conceived and guided the research: Noel Ortuño. Performed the laboratory experiments: José Antonio Castillo, Mayra Claros, Marlene Angulo, Violeta Angulo. Performed greenhouse assays: Mayra Claros, Marlene Angulo. Performed field trials: Oscar Navia. Assayed for large-scale multiplication of microorganisms and formulation: Daniel Barja, Claudia Gutiérrez. Wrote the paper: José Antonio Castillo.

Conflicts of Interest

The authors declare no conflict of interest.

REFERENCES

1. Food and Agriculture Organization of the United Nations. *A Provisional Methodology for Soil Degradation Assessment;* Food and Agriculture Organization: Rome, Italy, 1979; Volume 1, p. 84.
2. Wright, K.H.; Pike, O.A.; Fairbanks, D.J.; Huber, C.S. Composition of *Atriplex hortensis,* sweet and bitter *Chenopodium quinoa* seeds. *J. Food Sci.* **2002**, *67*,1383-1385.
3. Bolivian Institute of Foreign Trade, Bolivia: Exportaciones de Quinua, CIFRAS, Biweekly Newsletter N° 195, IBCE, Santa Cruz, Bolivia. Available online: http://ibce.org.bo/publicaciones-ibcecifras.php?op=1&id=123 (accessed on June 14, 2013).
4. Benizri, E.; Baudoin, E.; Guckert, A. Root colonization by inoculated plant growth promoting rhizobacteria. *Biocontrol. Sci. Technol.* **2001**, *11*,557-574.
5. Bacilio-Jimenez, M.; Aguilar-Flores, S.; Del Valle, M.V.; Perez, A; Zepeda, A; Zenteno, E. Endophytic bacteria in rice seeds inhibit early colonization of roots by *Azospirillum brasilense. Soil Biol. Biochem.* **2001**, *33*,167-172.
6. Cattelan, A.J.; Hartel, P.G.; Fuhrmann, J.J. Screening for plant growth-promoting rhizobacteria to promote soybean growth. *Soil Sci. Soc. Am. J.* **1999**, *63*, 1670-1680.
7. Fürnkranz, M.; Müller, H.; Berg, G. Characterization of plant growth promoting bacteria from crops in Bolivia. *J. Plant Dis. Protect.* **2009**, *116*, 149-155.
8. Timmusk, S. Mechanism of Action of the Plant Growth-Promoting Bacterium *Paenibacillus polymixa.* Ph.D. Thesis, Uppsala University, Uppsala, Sweden, 4 December 2003.
9. Bashan, Y.; Puente, M.E.; de-Bashan, L.E.; Hernandez, J.P. Environmental Uses of Plant Growth-Promoting Bacteria. In *Plant-Microbe Interactions;* Ait-Barka, E., Clément, C., Eds.; Research Signpost: Trivandrum, India, 2008; pp. 69–93.
10. Datnoff, L.E.; Pernezny, K.L. Effect of bacterial and fungal microorganisms to colonize tomato roots, improve transplant growth and control of Fusarium crown and root rot. *Proc. Florida Tomato Inst.* **1998**, *111*, 26–33.
11. Hermosa, R.; Viterbo, A.; Chet, I.; Monte, E. Plant-beneficial effects of *Trichoderma* and of its genes. *Microbiology* **2012**, *158*, 17–25.
12. Ortuño, N.; Claros, M. Fundación Proinpa, Cochabamba, Bolivia. Unpublished work, 2010.
13. Nautical, C.S. An efficient microbiological growth medium for screening phosphate solubilizing microorganisms. *FEMS Microbiol. Lett.* **1999**, *170*, 265–270.
14. Gordon, S.; Weber, R. Colorimetric estimation of indole acetic acid. *Plant Physiol.* **1951**, *26*, 192–195.
15. Wilson, K. Preparation of genomic DNA from bacteria. In *Current Protocols in Molecular Biology;* Ausubel, F.M., Brent, R., Kingston, R.E., Moore, D.D., Seidman, J.G., Smith, J.A., Struhl, K., Eds.; John Wiley & Sons: New York, NY, USA, 1994; pp. 241–245.
16. Melo, S.C.O.; Pungartnik, C.; Cascardo, J.C.M.; Brendel, M. Rapid and efficient protocol for DNA extraction and molecular identification of the basidiomycete *Crinipellis perniciosa. Genet. Mol. Res.* **2006**, *5*, 851–855.
17. Hall, T.A. BioEdit: A user-friendly biological sequence alignment editor and analysis program for Windows 95/98/NT. *Nucleic Acids Symp. Ser.* **1999**, *41*, 95–98.
18. Altschul, S.F.; Gish, W.; Miller, W.; Myers, E.W.; Lipman, D.J. Basic local alignment search tool. *J. Mol. Biol.* **1990**, *215*, 403–410.

19. International Subcommission on Trichoderma and Hypocrea website. Available online: www.isth.info (accessed on 23 July 2012).

20. Druzhinina, I.; Kopchinskiy, A.G.; Komon, M.; Bissett, J.; Szakacs, G.; Kubicek, C.P. An oligonucleotide barcode for species identification in *Trichoderma* and *Hypocrea*. *Fungal. Genet. Biol.* **2005**, *42*, 813–828.

21. Kopchinskiy, A.; Komon, M.; Kubicek, C.P.; Druzhinina, I.S. TrichoBLAST: A multiloci database of phylogenetic markers for *Trichoderma* and *Hypocrea* powered by sequence diagnosis and similarity search tools. *Mycol. Res.* **2005**, *109*, 658–660.

22. Lane, D.J. 16S/23S Sequencing. In *Nucleic Acid Techniques in Bacterial Systematics*; Stackebrandt, E., Goodfellow, M., Eds.; John Wiley & Sons: Chichester, UK, 1991; pp. 115–175.

23. Gardes, M.; Bruns, T.D. ITS primers with enhanced specificity for basidiomycetes—Application to the identification of mycorrhizae and rusts. *Mol. Ecol.* **1993**, *2*, 113–118.

24. White, T.J.; Bruns, T.; Lee, S.; Taylor, J. Amplification and direct sequencing of fungal ribosomal RNA genes for phylogenetics. In *PCR Protocols: A Guide to Methods and Applications*; Innis, M.A., Gelfand, D.H., Sninsky, J.J., White, T.J., Eds.; Academic Press: San Diego, CA, USA, 1990; pp. 315–322.

25. Chaverri, P.; Castlebury, L.A.; Overton, B.E.; Samuels, G.J. *Hypocrea/Trichoderma*: Species with conidiophore elongations and green conidia. *Mycologia* **2003**, *95*, 1100–1140.

26. Nagy, V.; Seidl, V.; Szakacs, G.; Komon-Zelazowska, M.; Kubicek, C.P.; Druzhinina, I.S. Application of DNA bar codes for screening of industrially important fungi: The haplotype of *Trichoderma harzianum* sensu stricto indicates superior chitinase formation. *Appl. Environ. Microbiol.* **2007**, *73*, 7048–7058.

27. Bartholomew, J.W.; Mittwer, T. A simplified bacterial spore stain. *Stain Technol.* **1950**, *25*, 153–156.

This page left intentionally blank.

INDEX

D

E

F

G